*Report Of
The President's Commission On*
THE ACCIDENT AT THREE MILE ISLAND

*The Need For Change:
The Legacy Of TMI
October 1979 Washington, D.C.*

Pergamon Press

New York ☐ Oxford ☐ Toronto ☐ Sydney ☐ Frankfurt ☐ Paris

Pergamon Press Offices:

U.S.A.	Pergamon Press Inc., Maxwell House, Fairview Park, Elmsford, New York 10523, U.S.A.
U.K.	Pergamon Press Ltd., Headington Hill Hall, Oxford OX3 0BW, England
CANADA	Pergamon of Canada, Ltd., 150 Consumers Road, Willowdale, Ontario M2J, 1P9, Canada
AUSTRALIA	Pergamon Press (Aust) Pty. Ltd., P O Box 544, Potts Point, NSW 2011, Australia
FRANCE	Pergamon Press SARL, 24 rue des Ecoles, 75240 Paris, Cedex 05, France
FEDERAL REPUBLIC OF GERMANY	Pergamon Press GmbH, 6242 Kronberg/Taunus, Pferdstrasse 1, Federal Republic of Germany

ISBN-0-08-025946-4

This edition is a reprint of *Report of the President's Commission on The Accident at Three Mile Island.*

Printed in the United States of America

The President's Commission On
THE ACCIDENT AT TMI

John G. Kemeny, Chairman
President, Dartmouth College

Bruce Babbitt
  Governor of Arizona
Patrick E. Haggerty
  Honorary Chairman and
  General Director
  Texas Instruments Incorporated
Carolyn Lewis
  Associate Professor
  Graduate School of Journalism
  Columbia University
Paul A. Marks
  Vice President for Health Sciences
  and Frode Jensen Professor
  Columbia University
Cora B. Marrett
  Professor of Sociology and
  Afro-American Studies
  University of Wisconsin-Madison
Lloyd McBride
  President
  United Steelworkers of America

Harry C. McPherson
  Partner
  Verner, Liipfert,
  Bernhard, and McPherson
Russell W. Peterson
  President
  National Audubon Society
Thomas H. Pigford
  Professor and Chairman
  Department of Nuclear
  Engineering
  University of California at
  Berkeley
Theodore B. Taylor
  Visiting Lecturer
  Department of Mechanical
  and Aerospace Engineering
  Princeton University
Anne D. Trunk
  Resident
  Middletown, Pennsylvania

Senior Professional Staff

Stanley M. Gorinson, Chief Counsel

Kevin P. Kane, Deputy Chief Counsel

Associate Chief Counsel
 Charles A. Harvey, Jr.
 Stan M. Helfman
 Winthrop A. Rockwell

Vincent L. Johnson, Director of Technical Staff

Task Force Heads
 Russell R. Dynes
 Jacob I. Fabrikant
 Leonard Jaffe
 David M. Rubin

Francis T. Hoban, Director of Administration

Barbara Jorgenson, Public Information Director

Patrick Young, Senior Writer

Consultants to the Chairman
 Harold Bruff
 Bruce T. Lundin
 Jonathan Moore

TABLE OF CONTENTS

Preface 1

Overview 7

Commission Findings 27

Commission Recommendations 61

Account of the Accident 81

Appendices

Executive Order 151

Commission Operations & Methodology 155

Commissioners' Biographies 157

Staff List 165

Glossary 173

Supplemental View by Six Commisioners 181

**President's Commission
on the Accident at Three Mile Island**
2100 M Street, NW Washington, DC 20037

October 30, 1979

The President
The White House
Washington, D.C. 20500

Dear Mr. President:

    In accordance with Executive Order Number 12130, we hereby transmit to you the final report of the President's Commission on the Accident at Three Mile Island.

                            Faithfully yours,

                            John G. Kemeny
                            Chairman

Bruce Babbitt   Harry C. McPherson

Patrick E. Haggerty   Russell W. Peterson

Carolyn Lewis   Thomas H. Pigford

Paul A. Marks   Theodore B. Taylor

Cora B. Marrett   Anne D. Trunk

Lloyd McBride

PREFACE

THE CHARGE TO THE COMMISSION

On March 28, 1979, the United States experienced the worst accident in the history of commercial nuclear power generation. Two weeks later, the President of the United States established a Presidential Commission. The President charged the 12-member Commission as follows:

"The purpose of the Commission is to conduct a comprehensive study and investigation of the recent accident involving the nuclear power facility on Three Mile Island in Pennsylvania. The Commission's study and investigation shall include:

> (a) a technical assessment of the events and their causes; this assessment shall include, but shall not be limited to, an evaluation of the actual and potential impact of the events on the public health and safety and on the health and safety of workers;
>
> (b) an analysis of the role of the managing utility;
>
> (c) an assessment of the emergency preparedness and response of the Nuclear Regulatory Commission and other federal, state, and local authorities;
>
> (d) an evaluation of the Nuclear Regulatory Commission's licensing, inspection, operation, and enforcement procedures as applied to this facility;
>
> (e) an assessment of how the public's right to information concerning the events at TMI was served and of the steps which should be taken during similar emergencies to provide the public with accurate, comprehensible, and timely information; and
>
> (f) appropriate recommendations based upon the Commission's findings."

PREFACE

THE ACCIDENT

At 4:00 a.m. on March 28, 1979, a serious accident occurred at the Three Mile Island 2 nuclear power plant near Middletown, Pennsylvania. The accident was initiated by mechanical malfunctions in the plant and made much worse by a combination of human errors in responding to it. (For details see "Account of the Accident" within this volume.) During the next 4 days, the extent and gravity of the accident was unclear to the managers of the plant, to federal and state officials, and to the general public. What is quite clear is that its impact, nationally and internationally, has raised serious concerns about the safety of nuclear power. This Commission was established in response to those concerns.

WHAT WE DID

The investigation of the Commission was carried out by our able and hard-working staff. We also had the help of a number of consultants and commissioned several studies. It is primarily due to the work of the staff that we accomplished the following.

We examined with great care the sequence of events that occurred during the accident, to determine what happened and why. We have attempted to evaluate the significance of various equipment failures as well as the importance of actions (or failures of actions) on the parts of individuals and organizations.

We analyzed the various radiation releases and came up with the best possible estimates of the health effects of the accident. In addition, we looked more broadly into how well the health and safety of the workers was protected during normal operating conditions, and how well their health and safety and that of the general public would have been protected in the case of a more serious accident.

We conducted an in-depth examination of the role played by the utility and its principal suppliers. We examined possible problems of organization, procedures, and practices that might have contributed to the accident. Since the major cause of the accident was due to inappropriate actions by those who were operating the plant and supervising that operation, we looked very carefully at the training programs that prepare operators and the procedures under which they operate.

As requested by the President, we examined the emergency plans that were in place at the time of the accident. We also probed the responses to the accident by the utility, by state and local governmental agencies in Pennsylvania, and by a variety of federal agencies. We looked for deficiencies in the plans and in their execution in order to be able to make recommendations for improvements for any future accident. In this process we had in mind how well the response would have worked if the danger to public health had been significantly greater.

We examined the coverage of the accident by the news media. This was a complex process in which we had to separate out whether errors in media accounts were due to ignorance or confusion on the part of the official sources, to the way they communicated this information to the

PREFACE

media, or to mistakes committed by the reporters themselves. We examined what sources were most influential on the people who needed immediate information, and how well the public was served by the abundant coverage that was provided. We also attempted to evaluate whether the coverage tended to exaggerate the seriousness of the accident either by selectively using alarming quotes more than reassuring ones, or through purposeful sensationalism.

Finally, we spent a great deal of time on the agency that had a major role in all of the above: the Nuclear Regulatory Commission. The President gave us a very broad charge concerning this agency. We therefore tried to understand its complex structure and how well it functions, its role in licensing and rulemaking, how well it carries out its mission through its inspection and enforcement program, the role it plays in monitoring the training of operators, and its participation in the response to the emergency, including the part it played in providing information to the public.

We took more than 150 formal depositions and interviewed a significantly larger number of individuals. At our public hearings we heard testimony under oath from a wide variety of witnesses. We collected voluminous material that will fill about 300 feet of shelf-space in a library. All of this material will be placed into the National Archives. The most important information extracted from this in each of the areas will appear in a series of "Staff Reports to the Commission."

Based on all of this information, the Commission arrived at a number of major findings and conclusions. In turn, these findings led the Commission to a series of recommendations responsive to the President's charge.

At the beginning of this volume will be found an overview of our investigation, followed by those findings and recommendations which commanded a significant consensus among the members of the Commission. Each recommendation was approved by a majority of Commissioners.

WHAT WE DID NOT DO

It is just as important for the reader to understand what the Commission did not do.

Our investigation centered on one accident at one nuclear power plant in the United States. While acting under the President's charge, we had to look at a large number of issues affecting many different organizations; there are vast related issues which were outside our charge, and which we could not possibly have examined in a 6-month investigation.

We did not examine the entire nuclear industry. (Although, through our investigation of the Nuclear Regulatory Commission, we have at least some idea of the standards being applied to it across the board.) We have not looked at the military applications of nuclear energy. We did not consider nuclear weapons proliferation. We have not dealt with the question of the disposal of radioactive waste or the dangers of the accumulation of waste fuel within nuclear power plants adjacent to the

PREFACE

containment buildings. We made no attempt to examine the entire fuel cycle, starting with the mining of uranium. And, of course, we made no examination of the many other sources of radiation, both natural and man-made, that affect all of us.

We have not attempted to evaluate the relative risks involved in alternate sources of energy. We are aware of a number of studies that try to do this. We are also aware that some of these studies are subjects of continuing controversy.

We did not attempt to reach a conclusion as to whether, as a matter of public policy, the development of commercial nuclear power should be continued or should not be continued. That would require a much broader investigation, involving economic, environmental, and political considerations. We are aware that there are 72 operating reactors in the United States with a capacity of 52,000 megawatts of electric energy. An additional 92 plants have received construction permits and are in various stages of construction. If these are completed, they will roughly triple the present nuclear capacity to generate electricity. This would be a significant fraction of the total U.S. electrical generating capacity of some 600,000 megawatts. In addition, there are about 200 nuclear power plants in other countries throughout the world.

Therefore, the improvement of the safety of existing and planned nuclear power plants is a crucial issue. It is this issue that our report addresses, those changes that can and must be made as a result of the accident -- the legacy of Three Mile Island.

OVERVIEW

OVERALL CONCLUSION

In announcing the formation of the Commission, the President of the United States said that the Commission "will make recommendations to enable us to prevent any future nuclear accidents." After a 6-month investigation of all factors surrounding the accident and contributing to it, the Commission has concluded that:

<u>To prevent nuclear accidents as serious as Three Mile Island, fundamental changes will be necessary in the organization, procedures, and practices -- and above all -- in the attitudes of the Nuclear Regulatory Commission and, to the extent that the institutions we investigated are typical, of the nuclear industry.</u>

This conclusion speaks of <u>necessary</u> fundamental changes. We do not claim that our proposed recommendations are sufficient to assure the safety of nuclear power.

Given the nature of its Presidential mandate, its time limitations, and the complexity of both energy and comparative "risk-assessment" issues, this Commission has not undertaken to examine how safe is "safe enough" or the broader question of nuclear versus other forms of energy. The Commission's findings with respect to the accident and the regulation of the nuclear industry -- particularly the current and potential state of public safety in the presence of nuclear power -- have, we believe, implications that bear on the broad question of energy. But the ultimate resolution of the question involves the kind of economic, environmental, and foreign policy considerations that can only be evaluated through the political process.

Our findings do not, standing alone, require the conclusion that nuclear power is inherently too dangerous to permit it to continue and expand as a form of power generation. Neither do they suggest that the nation should move forward aggressively to develop additional commercial nuclear power. They simply state that if the country wishes, for larger reasons, to confront the risks that are inherently associated with

OVERVIEW

nuclear power, fundamental changes are necessary if those risks are to be kept within tolerable limits.

We are very much aware that many other investigations into the accident are under way. There are several investigations by Congress, the NRC self-investigation, and a number of studies by the industry. Some will examine individual issues in much greater depth than we were able to do. And, no doubt, additional insights will emerge out of these various investigations. It is our hope that the results of our efforts may aid and accelerate the progress of the ongoing investigations, and help to bring about the required changes promptly.

ATTITUDES AND PRACTICES

Our investigation started out with an examination of the accident at Three Mile Island (TMI). This necessarily led us to look into the role played by the utility and its principal suppliers. With our in-depth investigation of the Nuclear Regulatory Commission (NRC), we gained a broader insight into the attitudes and practices that prevail in portions of the industry. However, we did not examine the industry in its totality.

Popular discussions of nuclear power plants tend to concentrate on questions of equipment safety. Equipment can and should be improved to add further safety to nuclear power plants, and some of our recommendations deal with this subject. But as the evidence accumulated, it became clear that the fundamental problems are people-related problems and not equipment problems.

When we say that the basic problems are people-related, we do not mean to limit this term to shortcomings of individual human beings -- although those do exist. We mean more generally that our investigation has revealed problems with the "system" that manufactures, operates, and regulates nuclear power plants. There are structural problems in the various organizations, there are deficiencies in various processes, and there is a lack of communication among key individuals and groups.

We are convinced that if the only problems were equipment problems, this Presidential Commission would never have been created. The equipment was sufficiently good that, except for human failures, the major accident at Three Mile Island would have been a minor incident. But, wherever we looked, we found problems with the human beings who operate the plant, with the management that runs the key organization, and with the agency that is charged with assuring the safety of nuclear power plants.

In the testimony we received, one word occurred over and over again. That word is "mindset." At one of our public hearings, Roger Mattson, director of NRC's Division of Systems Safety, used that word five times within a span of 10 minutes. For example: "I think [the] mindset [was] that the operator was a force for good, that if you discounted him, it was a measure of conservatism." In other words, they

OVERVIEW

concentrated on equipment, assuming that the presence of operators could only improve the situation -- they would not be part of the problem.

After many years of operation of nuclear power plants, with no evidence that any member of the general public has been hurt, the belief that nuclear power plants are sufficiently safe grew into a conviction. One must recognize this to understand why many key steps that could have prevented the accident at Three Mile Island were not taken. The Commission is convinced that this attitude must be changed to one that says nuclear power is by its very nature potentially dangerous, and, therefore, one must continually question whether the safeguards already in place are sufficient to prevent major accidents. A comprehensive system is required in which equipment and human beings are treated with equal importance.

We note a preoccupation with regulations. It is, of course, the responsibility of the Nuclear Regulatory Commission to issue regulations to assure the safety of nuclear power plants. However, we are convinced that regulations alone cannot assure safety. Indeed, once regulations become as voluminous and complex as those regulations now in place, they can serve as a negative factor in nuclear safety. The regulations are so complex that immense efforts are required by the utility, by its suppliers, and by the NRC to assure that regulations are complied with. The satisfaction of regulatory requirements is equated with safety. This Commission believes that it is an absorbing concern with safety that will bring about safety -- not just the meeting of narrowly prescribed and complex regulations.

We find a fundamental fault even with the existing body of regulations. While scientists and engineers have worried for decades about the safety of nuclear equipment, we find that the approach to nuclear safety had a major flaw. It was natural for the regulators and the industry to ask: "What is the worst kind of equipment failure that can occur?" Some potentially serious scenarios, such as the break of a huge pipe that carries the water cooling the nuclear reactor, were studied extensively and diligently, and were used as a basis for the design of plants. A preoccupation developed with such large-break accidents as did the attitude that if they could be controlled, we need not worry about the analysis of "less important" accidents.

Large-break accidents require extremely fast reaction, which therefore must be automatically performed by the equipment. Lesser accidents may develop much more slowly and their control may be dependent on the appropriate actions of human beings. This was the tragedy of Three Mile Island, where the equipment failures in the accident were significantly less dramatic than those that had been thoroughly analyzed, but where the results confused those who managed the accident. A potentially insignificant incident grew into the TMI accident, with severe damage to the reactor. Since such combinations of minor equipment failures are likely to occur much more often than the huge accidents, they deserve extensive and thorough study. In addition, they require operators and supervisors who have a thorough understanding of the functioning of the plant and who can respond to combinations of small equipment failures.

OVERVIEW

The most serious "mindset" is the preoccupation of everyone with the safety of equipment, resulting in the down-playing of the importance of the human element in nuclear power generation. We are tempted to say that while an enormous effort was expended to assure that safety-related equipment functioned as well as possible, and that there was backup equipment in depth, what the NRC and the industry have failed to recognize sufficiently is that the human beings who manage and operate the plants constitute an important safety system.

CAUSES OF THE ACCIDENT

Other investigations have concluded that, while equipment failures initiated the event, the fundamental cause of the accident was "operator error." It is pointed out that if the operators (or those who supervised them) had kept the emergency cooling systems on through the early stages of the accident, Three Mile Island would have been limited to a relatively insignificant incident. While we agree that this statement is true, we also feel that it does not speak to the fundamental causes of the accident.

Let us consider some of the factors that significantly contributed to operator confusion.

First of all, it is our conclusion that the training of TMI operators was greatly deficient. While training may have been adequate for the operation of a plant under normal circumstances, insufficient attention was paid to possible serious accidents. And the depth of understanding, even of senior reactor operators, left them unprepared to deal with something as confusing as the circumstances in which they found themselves.

Second, we found that the specific operating procedures, which were applicable to this accident, are at least very confusing and could be read in such a way as to lead the operators to take the incorrect actions they did.

Third, the lessons from previous accidents did not result in new, clear instructions being passed on to the operators. Both points are illustrated in the following case history.

A senior engineer of the Babcock & Wilcox Company (suppliers of the nuclear steam system) noted in an earlier accident, bearing strong similarities to the one at Three Mile Island, that operators had mistakenly turned off the emergency cooling system. He pointed out that we were lucky that the circumstances under which this error was committed did not lead to a serious accident and warned that under other circumstances (like those that would later exist at Three Mile Island), a very serious accident could result. He urged, in the strongest terms, that clear instructions be passed on to the operators. This memorandum was written 13 months before the accident at Three Mile Island, but no new instructions resulted from it. The Commission's investigation of this incident, and other similar incidents within B&W and the NRC, indicates that the lack of understanding that led the operators to

OVERVIEW

incorrect action existed both within the Nuclear Regulatory Commission and within the utility and its suppliers.

We find that there is a lack of "closure" in the system -- that is, important safety issues are frequently raised and may be studied to some degree of depth, but are not carried through to resolution; and the lessons learned from these studies do not reach those individuals and agencies that most need to know about them. This was true in the B&W incident described above, it was true about various warnings within NRC that inappropriate operator actions could result in the case of certain small-break accidents, and it was true in several examples of questions raised in connection with licensing procedures that were not followed to their conclusion by the NRC staff.

There are many other examples mentioned in our report that indicate the lack of attention to the human factor in nuclear safety. We note only one more (a fourth) example. The control room, through which the operation of the TMI-2 plant is carried out, is lacking in many ways. The control panel is huge, with hundreds of alarms, and there are some key indicators placed in locations where the operators cannot see them. There is little evidence of the impact of modern information technology within the control room. In spite of this, this control room might be adequate for the normal operation of nuclear power plants.

However, it is seriously deficient under accident conditions. During the first few minutes of the accident, more than 100 alarms went off, and there was no system for suppressing the unimportant signals so that operators could concentrate on the significant alarms. Information was not presented in a clear and sufficiently understandable form; for example, although the pressure and temperature within the reactor coolant system were shown, there was no direct indication that the combination of pressure and temperature meant that the cooling water was turning into steam. Overall, little attention had been paid to the interaction between human beings and machines under the rapidly changing and confusing circumstances of an accident. Perhaps these design failures were due to a concentration on the large-break accidents -- which do not allow time for significant operator action -- and the design ignored the needs of operators during a slowly developing small-break (TMI-type) accident. While some of us may favor a complete modernization of control rooms, we are all agreed that a relatively few and not very expensive improvements in the control room could have significantly facilitated the management of the accident.

In conclusion, while the major factor that turned this incident into a serious accident was inappropriate operator action, many factors contributed to the action of the operators, such as deficiencies in their training, lack of clarity in their operating procedures, failure of organizations to learn the proper lessons from previous incidents, and deficiencies in the design of the control room. These shortcomings are attributable to the utility, to suppliers of equipment, and to the federal commission that regulates nuclear power. Therefore -- whether or not operator error "explains" this particular case -- given all the above deficiencies, we are convinced that an accident like Three Mile Island was eventually inevitable.

OVERVIEW

SEVERITY OF THE ACCIDENT

Just how serious was the accident? Based on our investigation of the health effects of the accident, we conclude that in spite of serious damage to the plant, most of the radiation was contained and the actual release will have a negligible effect on the physical health of individuals. The major health effect of the accident was found to be mental stress.

The amount of radiation received by any one individual outside the plant was very low. However, even low levels of radiation may result in the later development of cancer, genetic defects, or birth defects among children who are exposed in the womb. Since there is no direct way of measuring the danger of low-level radiation to health, the degree of danger must be estimated indirectly. Different scientists make different assumptions about how this estimate should be made and, therefore, estimates vary. Fortunately, in this case the radiation doses were so low that we conclude that the overall health effects will be minimal. There will either be no case of cancer or the number of cases will be so small that it will never be possible to detect them. The same conclusion applies to the other possible health effects. The reasons for these conclusions are as follows.

An example of a projection derived for the total number of radiation-induced cancers among the population affected by the accident at TMI was 0.7. This number is an estimate of an average, such as the one that appears in the statement: "The average American family has 2.3 children."

In the case of TMI, what it really means is that each of some 2 million individuals living within 50 miles has a miniscule additional chance of dying of cancer, and when all of these minute probabilities are added up, they total 0.7. In such a situation, a mathematical law known as a Poisson distribution (named after a famous French mathematician) applies. If the estimated average is 0.7, then the actual probabilities for cancer deaths due to the accident work out as follows: There is a roughly 50 percent chance that there will be no additional cancer deaths, a 35 percent chance that one individual will die of cancer, a 12 percent chance that two people will die of cancer, and it is practically certain that there will not be as many as five cancer deaths.

Similar probabilities can be calculated for our various estimates. All of them have in common the following: It is entirely possible that not a single extra cancer death will result. And for all our estimates, it is practically certain that the additional number of cancer deaths will be less than 10.

Since a cancer caused by nuclear radiation is no different from any other cancer, additional cancers can only be determined statistically. We know from statistics on cancer deaths that among the more than 2 million people living within 50 miles of TMI, eventually some 325,000 people will die of cancer, for reasons having nothing to do with the

OVERVIEW

nuclear power plant. Again, this number is only an estimate, and the actual figure could be as much as 1,000 higher or 1,000 lower. Therefore, there is no conceivable statistical method by which fewer than 10 additional deaths would ever be detected. Therefore, the accident may result in no additional cancer deaths or, if there were any, they would be so few that they could not be detected.

We found that the mental stress to which those living within the vicinity of Three Mile Island were subjected was quite severe. There were several factors that contributed to this stress. Throughout the first week of the accident, there was extensive speculation on just how serious the accident might turn out to be. At various times, senior officials of the NRC and the state government were considering the possibility of a major evacuation. There were a number of advisories recommending steps short of a full evacuation. Some significant fraction of the population in the immediate vicinity voluntarily left the region. NRC officials contributed to the raising of anxiety in the period from Friday to Sunday (March 30-April 1). On Friday, a mistaken interpretation of the release of a burst of radiation led some NRC officials to recommend immediate evacuation. And on Friday Governor Thornburgh advised pregnant women and preschool aged children within 5 miles of TMI to leave the area. On Saturday and Sunday, other NRC officials mistakenly believed that there was an imminent danger of an explosion of a hydrogen bubble within the reactor vessel, and evacuation was again a major subject of discussion.

We conclude that the most serious health effect of the accident was severe mental stress, which was short-lived. The highest levels of distress were found among those living within 5 miles of TMI and in families with preschool children.

There was very extensive damage to the plant. While the reactor itself has been brought to a "cold shutdown," there are vast amounts of radioactive material trapped within the containment and auxiliary buildings. The utility is therefore faced with a massive cleanup process that carries its own potential dangers to public health. The ongoing cleanup operation at TMI demonstrates that the plant was inadequately designed to cope with the cleanup of a damaged plant. The direct financial cost of the accident is enormous. Our best estimate puts it in a range of $1 to $2 billion, even if TMI-2 can be put back into operation. (The largest portion of this is for replacement power estimated for the next few years.) And since it may not be possible to put it back into operation, the cost could even be much larger.

The accident raised concerns all over the world and led to a lowering of public confidence in the nuclear industry and in the NRC.

From the beginning, we felt it important to determine not only how serious the actual impact of the accident was on public health, but whether we came close to a catastrophic accident in which a large number of people would have died. Issues that had to be examined were whether a chemical (hydrogen) or steam explosion could have ruptured the reactor vessel and containment building, and whether extremely hot molten fuel could have caused severe damage to the containment. The danger was never -- and could not have been -- that of a nuclear explosion (bomb).

We have made a conscientious effort to get an answer to this difficult question. Since the accident was due to a complex combination of minor equipment failures and major inappropriate human actions, we have asked the question: "What if one more thing had gone wrong?"

We explored each of several different scenarios representing a change in the sequence of events that actually took place. The greatest concern during the accident was that significant amounts of radioactive material (especially radioactive iodine) trapped within the plant might be released. Therefore, in each case, we asked whether the amount released would have been smaller or greater, and whether large amounts could have been released.

Some of these scenarios lead to a more favorable outcome than what actually happened. Several other scenarios lead to increases in the amount of radioactive iodine released, but still at levels that would not have presented a danger to public health. But we have also explored two or three scenarios whose precise consequences are much more difficult to calculate. They lead to more severe damage to the core, with additional melting of fuel in the hottest regions. These consequences are, surprisingly, independent of the age of the fuel.

Because of the uncertain physical condition of the fuel, cladding, and core, we have explored certain special and severe conditions that would, unequivocally, lead to a fuel-melting accident. In this sequence of events fuel melts, falls to the bottom of the vessel, melts through the steel reactor vessel, and finally, some fuel reaches the floor of the containment building below the reactor vessel where there is enough water to cover the molten fuel and remove some of the decay heat. To contain such an accident, it is necessary to continue removing decay heat for a period of many months.

At this stage we approach the limits of our engineering knowledge of the interactions of molten fuel, concrete, steel, and water, and even the best available calculations have a degree of uncertainty associated with them. Our calculations show that even if a meltdown occurred, there is a high probability that the containment building and the hard rock on which the TMI-2 containment building is built would have been able to prevent the escape of a large amount of radioactivity. These results derive from very careful calculations, which hold only insofar as our assumptions are valid. We cannot be absolutely certain of these results.

Some of the limits of this investigation were: (1) We have not examined possible consequences of operator error during or after the fuel melting process which might compromise the effectiveness of containment; (2) We have not examined the vulnerability of the various electrical and plumbing penetrations through the walls or the doorways for people and equipment; (3) The analysis was specific to the TMI-2 design and location (for example, the bedrock under the plant); (4) We

OVERVIEW

recognize that we have only explored a limited number of alternatives to the question "What if . . .?" and, others may come up with a plausible scenario whose results would have been even more serious.

We strongly urge that research be carried out promptly to identify and analyze the possible consequences of accidents leading to severe core damage. Such knowledge is essential for coping with results of future accidents. It may also indicate weaknesses in present designs, whose correction would be important for the prevention of serious accidents.

These uncertainties have not prevented us from reaching an overwhelming consensus on corrective measures. Our reasoning is as follows: Whether in this particular case we came close to a catastrophic accident or not, this accident was too serious. Accidents as serious as TMI should not be allowed to occur in the future.

The accident got sufficiently out of hand so that those attempting to control it were operating somewhat in the dark. While today the causes are well understood, 6 months after the accident it is still difficult to know the precise state of the core and what the conditions are inside the reactor building. Once an accident reaches this stage, one that goes beyond well-understood principles, and puts those controlling the accident into an experimental mode (this happened during the first day), the uncertainty of whether an accident could result in major releases of radioactivity is too high. Adding to this the enormous damage to the plant, the expensive and potentially dangerous cleanup process that remains, and the great cost of the accident, we must conclude that -- whatever worse could have happened -- the accident had already gone too far to make it tolerable.

While throughout this entire document we emphasize that fundamental changes are necessary to prevent accidents as serious as TMI, we must not assume that an accident of this or greater seriousness cannot happen again, even if the changes we recommend are made. Therefore, in addition to doing everything to prevent such accidents, we must be fully prepared to minimize the potential impact of such an accident on public health and safety, should one occur in the future.

HANDLING OF THE EMERGENCY

Another area of our investigation dealt with the questions of whether various agencies made adequate preparations for an emergency and whether their responses to the emergency were satisfactory. Our finding is negative on both questions.

We are disturbed both by the highly uneven quality of emergency plans and by the problems created by multiple jurisdictions in the case of a radiation emergency. Most emergency plans rely on prompt action at the local level to initiate a needed evacuation or to take other protective action. We found an almost total lack of detailed plans in the local communities around Three Mile Island. It is one of the many ironies of this event that the most relevant planning by local

OVERVIEW

authorities took place during the accident. In an accident in which prompt defensive steps are necessary within a matter of hours, insufficient advance planning could prove extremely dangerous.

We favor the centralization of emergency planning and response in a single agency at the federal level with close coordination between it and state and local agencies. Such agencies would need expert input from many other organizations, but there should be a single agency that has the responsibility both for assuring that adequate planning takes place and for taking charge of the response to the emergency. This will require organizational changes, since the agencies now best organized to deal with emergencies tend to have most of their experience with such events as floods and storms, rather than with radiological events. And, insofar as radiological events require steps that go beyond those in a normal emergency, careful additional planning is needed.

A central concept in the current siting policy of the NRC is that reactors should be located in a "low population zone" (LPZ), an area around the plant in which appropriate protective action could be taken for the residents in the event of an accident. However, this concept is implemented in a strange, unnatural, and round-about manner. To determine the size of the LPZ, the utility calculates the amount of radiation released in a very serious hypothetical accident. Using geographical and meteorological data, the utility then calculates that area within which an individual would receive 25,000 millirems or more to the whole body, during the entire course of the accident. This area is the LPZ. The 25,000-millirem standard is an extremely large dose, many times more serious than that received by any individual during the entire TMI accident.

The LPZ approach has serious shortcomings. First, because of the extremely large dose by which its size is determined, the LPZs for many nuclear power plants are relatively small areas, 2 miles in the case of TMI. Second, if an accident as serious as the one used to calculate the LPZ were actually to occur, it is evident that many people living outside the LPZ would receive smaller, but still massive doses of radiation. Third, the TMI accident shows that the LPZ has little relevance to the protection of the public -- the NRC itself was considering evacuation distances as far as 20 miles, even though the accident was far less serious than those postulated during siting. We have therefore concluded that the entire concept is flawed.

We recommend that the LPZ concept be abandoned in siting and in emergency planning. A variety of possible accidents should be considered during siting, particularly "smaller" accidents which have a higher probability of occurring. For each such accident, one should calculate probable levels of radiation releases at a variety of distances to decide the kinds of protective action that are necessary and feasible. Such protective actions may range from evacuation of an area near the plant, to the distribution of potassium iodide to protect the thyroid gland from radioactive iodine, to a simple instruction to people several miles from the plant to stay indoors for a specified period of time. Only such an analysis can predict the true consequences

OVERVIEW

of a radiological incident and determine whether a particular site is suitable for a nuclear power plant. Similarly, emergency plans should have built into them a variety of responses to a variety of possible kinds of accidents. State and local agencies must be prepared with the appropriate response once information is available on the nature of an accident and its likely levels of releases.

The response to the emergency was dominated by an atmosphere of almost total confusion. There was lack of communication at all levels. Many key recommendations were made by individuals who were not in possession of accurate information, and those who managed the accident were slow to realize the significance and implications of the events that had taken place. While we have attempted to address these shortcomings in our recommendations, it is important to reiterate the fundamental philosophy we stated above: One must do everything possible to prevent accidents of this seriousness, but at the same time assume that such an accident may occur and be prepared for response to the resulting emergency. The fact that too many individuals and organizations were not aware of the dimensions of serious accidents at nuclear power plants accounts for a great deal of the lack of preparedness and the poor quality of the response.

PUBLIC AND WORKER HEALTH AND SAFETY

We have identified a number of inadequacies with respect to procedures and programs to prevent or minimize hazards to health from radiation exposure from the operations of nuclear power plants. In setting standards for permissible levels of worker exposure to radioactivity, in plant siting decisions, and in other areas related to health, the NRC is not required to, and does not regularly seek, advice or review of its health-related guidelines and regulations from other federal agencies with radiation-related responsibilities in the area of health, for example the Department of Health, Education, and Welfare (HEW) or the Environmental Protection Agency (EPA). There is inadequate knowledge of the effects of low levels of ionizing radiation, of strategies to mitigate the health hazards of exposure to radiation, and of other areas relating to regulation setting to protect worker and public health. In preparation for a possible emergency such as the accident at TMI-2, various federal agencies (NRC, Department of Energy, HEW, and EPA) have assigned responsibilities, but planning prior to the accident was so poor that ad hoc arrangements among these federal agencies had to be made to involve them and coordinate their activities.

The Commonwealth of Pennsylvania, its Bureau of Radiation Protection and Department of Health -- agencies with responsibilities for public health -- did not have adequate resources for dealing with radiation health programs related to the operation of TMI. The utility was not required to, and did not, keep a record on workers of the total work-related plus non-work-related (for example, medical or dental) radiation exposure.

We make recommendations with respect to improving the coordination and collaboration among federal and state agencies with radiation-related

OVERVIEW

responsibilities in the health area. We believe more emphasis is required on research on the health effects of radiation to provide a sounder basis for guidelines and regulations related to worker and public health and safety. We believe that both the state and the utility have an opportunity and an obligation to establish more rigorous programs for informing workers and the public on radiation health-related issues and procedures to prevent adverse health effects of radiation.

RIGHT TO INFORMATION

The President asked us to investigate whether the public's right to information during the emergency was well served. Our conclusion is again in the negative. However, here there were many different causes, and it is both harder to assign proper responsibility and more difficult to come up with appropriate recommendations. There were serious problems with the sources of information, with how this information was conveyed to the press, and also with the way the press reported what it heard.

We do not find that there was a systematic attempt at a "cover-up" by the sources of information. Some of the official news sources were themselves confused about the facts and there were major disagreements among officials. On the first day of the accident, there was an attempt by the utility to minimize its significance, in spite of substantial evidence that it was serious. Later that week, NRC was the source of exaggerated stories. Due to misinformation, and in one case (the hydrogen bubble) through the commission of scientific errors, official sources would make statements about radiation already released (or about the imminent likelihood of releases of major amounts of radiation) that were not justified by the facts -- at least not if the facts had been correctly understood. And NRC was slow in confirming good news about the hydrogen bubble. On the other hand, the estimated extent of the damage to the core was not fully revealed to the public.

A second set of problems arose from the manner in which the facts were presented to the press. Some of those who briefed the press lacked the technical expertise to explain the events and seemed to be cut off from those who could have provided this expertise. When those who did have the knowledge spoke, their statements were often couched in "jargon" that was very difficult for the press to understand. The press was further disturbed by the fact that, in order to cut down on the amount of confusion, a number of potential sources of information were instructed not to give out information. While this cut down on the amount of confusion, it flew in the face of the long tradition of the press of checking facts with multiple sources.

Many factors contributed to making this event one of the most heavily covered media events ever. Given these circumstances, the media generally attempted to give a balanced presentation which would not contribute to an escalation of panic. There were, however, a few notable examples of irresponsible reporting and some of the visual images used in the reporting tended to be sensational.

OVERVIEW

Another severe problem was that even personnel representing the major national news media often did not have sufficient scientific and engineering background to understand thoroughly what they heard, and did not have available to them people to explain the information. This problem was most serious in the reporting of the various releases of radiation and the explanation of the severity (or lack of severity) of these releases. Many of the stories were so garbled as to make them useless as a source of information.

We therefore conclude that, while the extent of the coverage was justified, a combination of confusion and weakness in the sources of information and lack of understanding on the part of the media resulted in the public being poorly served.

In considering the handling of information during the nuclear accident, it is vitally important to remember the fear with respect to nuclear energy that exists in many human beings. The first application of nuclear energy was to atomic bombs which destroyed two major Japanese cities. The fear of radiation has been with us ever since and is made worse by the fact that, unlike floods or tornadoes, we can neither hear nor see nor smell radiation. Therefore, utilities engaged in the operation of nuclear power plants, and news media that may cover a possible nuclear accident, must make extraordinary preparation for the accurate and sensitive handling of information.

There is a natural conflict between the public's right to know and the need of disaster managers to concentrate on their vital tasks without distractions. There is no simple resolution for this conflict. But significant advance preparation can alleviate the problem. It is our judgment that in this case, neither the utility nor the NRC nor the media were sufficiently prepared to serve the public well.

THE NUCLEAR REGULATORY COMMISSION

We had a broad mandate from the President to investigate the Nuclear Regulatory Commission. When NRC was split off from the old Atomic Energy Commission, the purpose of the split was to separate the regulators from those who were promoting the peaceful uses of atomic energy. We recognize that the NRC has an assignment that would be difficult under any circumstances. But, we have seen evidence that some of the old promotional philosophy still influences the regulatory practices of the NRC. While some compromises between the needs of safety and the needs of an industry are inevitable, the evidence suggests that the NRC has sometimes erred on the side of the industry's convenience rather than carrying out its primary mission of assuring safety.

Two of the most important activities of NRC are its licensing function and its inspection and enforcement (I&E) activities. We found serious inadequacies in both.

In the licensing process, applications are only <u>required</u> to analyze "single-failure" accidents. They are not required to analyze what

OVERVIEW

happens when two systems fail independently of each other, such as the event that took place at TMI. There is a sharp delineation between those components in systems that are "safety-related" and those that are not. Strict reviews and requirements apply to the former; the latter are exempt from most requirements -- even though they can have an effect on the safety of the plant. We feel that this sharp either/or definition is inappropriate. Instead, there should be a system of priorities as to how significant various components and systems are for the overall safety of the plant. There seems to be a persistent assumption that plants can be made sufficiently safe to be "people-proof." Thus, not enough attention is paid to the training of operating personnel and operator procedures in the licensing process. And, finally, plants can receive an operating license with several safety issues still unresolved. This places such a plant into a regulatory "limbo" with jurisdiction divided between two different offices within NRC. TMI-2 was in this status at the time of the accident, 13 months after it received its operating license.

NRC's primary focus is on licensing and insufficient attention has been paid to the ongoing process of assuring nuclear safety. An important example of this is the case of "generic problems," that is, problems that apply to a number of different nuclear power plants. Once an issue is labeled "generic," the individual plant being licensed is not responsible for resolving the issue prior to licensing. That, in itself, would be acceptable, if there were a strict procedure within NRC to assure the timely resolution of generic problems, either by its own research staff, or by the utility and its suppliers. However, the evidence indicates that labeling of a problem as "generic" may provide a convenient way of postponing decision on a difficult question.

The old AEC attitude is also evident in reluctance to apply new safety standards to previously licensed plants. While we would accept a need for reasonable timetables for "backfitting," we did not find evidence that the need for improvement of older plants was systematically considered prior to Three Mile Island.

The existence of a vast body of regulations by NRC tends to focus industry attention narrowly on the meeting of regulations rather than on a systematic concern for safety. Furthermore, the nature of some of the regulations, in combination with the way rate bases are established for utilities, may in some instances have served as a deterrent for utilities or their suppliers to take the initiative in proposing measures for improved safety.

Previous studies of I&E have criticized this branch severely. Inspectors frequently fail to make independent evaluations or inspections. The manual according to which inspectors are supposed to operate is so voluminous that many inspectors do not understand precisely what they are supposed to do. There have been a number of incidents in which inspectors have had difficulty in getting their superiors to concentrate on serious safety issues. The analysis of reported incidents by licensees has tended to concentrate on equipment malfunction, and serious operator errors have not been focused on.

OVERVIEW

Finally, while the statutory authority to impose fines is fairly limited, a previous study shows that I&E has made minimal use of even this authority.

Since in many cases NRC does not have the first-hand information necessary to enforce its regulations, it must rely heavily on the industry's own records for its inspection and enforcement activities. NRC accumulates vast amounts of information on the operating experience of plants. However, prior to the accident there was no <u>systematic</u> method of evaluating these experiences, and no systematic attempt to look for patterns that could serve as a warning of a basic problem.

NRC is vulnerable to the charge that it is heavily equipment-oriented, rather than people-oriented. Evidence for this exists in the weak and understaffed branch of NRC that monitors operator training, in the fact that inspectors who investigate accidents concentrate on what went wrong with the equipment and not on what operators may have done incorrectly, in the lack of attention to the quality of procedures provided for operators, and in an almost total lack of attention to the interaction between human beings and machines.

In addition to all the other problems with the NRC, we are extremely critical of the role the organization played in the response to the accident. There was a serious lack of communication among the commissioners, those who were attempting to make the decisions about the accident in Bethesda, the field offices, and those actually on site. This lack of communication contributed to the confusion of the accident. We are also skeptical whether the collegial mode of the five commissioners makes them a suitable body for the management of an emergency, and of the agency itself.

We found serious managerial problems within the organization. These problems start at the very top. It is not clear to us what the precise role of the five NRC commissioners is, and we have evidence that they themselves are not clear on what their role should be. The huge bureaucracy under the commissioners is highly compartmentalized with insufficient communication among the major offices. We do not see evidence of effective managerial guidance from the top, and we do see evidence of some of the old AEC promotional philosophy in key officers below the top. The management problems have been made much harder by adoption of strict rules that prohibit the commissioners from talking with some of their key staff on issues involved in the licensing process; we believe that these rules have been applied in an unnecessarily severe form within this particular agency. The geographic spread, which places top management in Washington and most of the staff in Bethesda and Silver Spring, Maryland (and in other parts of the country), also inhibits the easy exchange of ideas.

We therefore conclude that there is no well-thought-out, integrated system for the assurance of nuclear safety within the current NRC.

We have found evidence of repeated in-depth studies and criticisms both from within the agency and from without, but we found very little

OVERVIEW

evidence that these studies have resulted in significant improvement. This fact gives us particular concern for the future of the present NRC.

For all these reasons we recommend a total restructuring of the NRC. We recommend that it be an independent agency within the executive branch, headed by a single administrator, who is in every sense chief executive officer, to be chosen from outside NRC. The new administrator must be provided with the freedom to reorganize and to bring new blood into the restructured NRC's staff. This new blood could result in the change of attitudes that is vital for the solution of the problems of the nuclear industry.

We have also recommended a number of other organizational and procedural changes designed to make the new agency truly effective in assuring the safety of nuclear power plants. Included in these are an oversight committee to monitor the performance of the restructured NRC and mandatory review by HEW of radiation-related health issues.

THE UTILITY

When the decision was made to make nuclear power available for the commercial generation of energy, it was placed into the hands of the existing electric utilities. Nuclear power requires management qualifications and attitudes of a very special character as well as an extensive support system of scientists and engineers. We feel that insufficient attention was paid to this by the General Public Utilities Corporation (GPU).

There is a divided system of decision-making within GPU and its subsidiaries. While the utility has legal responsibility for a wide range of fundamental decisions, from plant design to operator training, some utilities have to rely heavily on the expertise of their suppliers and on the Nuclear Regulatory Commission. Our report contains a number of examples where this divided responsibility, in the case of TMI, may have led to less than optimal design and operating practices. For example, we have received contradictory testimony on how the criteria under which the containment building isolates were selected. Similarly, the design of the control room seems to have been a compromise among of the utility, its parent company, the architect-engineer, and the nuclear steam system supplier (with very little attention from the NRC). But the clearest example of the shortcomings of divided responsibility is the area of operator training.

The legal responsibility for training operators and supervisors for safe operation of nuclear power plants rests with the utility. However, Met Ed, the GPU subsidiary which operates TMI, did not have sufficient expertise to carry out this training program without outside help. They, therefore, contracted with Babcock & Wilcox, supplier of the nuclear steam system, for various portions of this training program. While B&W has substantial expertise, they had no responsibility for the quality of the _total_ training program, only for carrying out the contracted portion. And coordination between the training programs of the two companies was extremely loose. For example, the B&W instructors were not aware of the precise operating procedures in effect at the plant.

OVERVIEW

A key tool in the B&W training is a "simulator," which is a mock control console that can reproduce realistically events that happen within a power plant. The simulator differs in certain significant ways from the actual control console. Also, the simulator was not programmed, prior to March 28, to reproduce the conditions that confronted the operators during the accident.

We found that at both companies, those most knowledgeable about the workings of the nuclear power plant have little communication with those responsible for operator training, and therefore, the content of the instructional program does not lead to sufficient understanding of reactor systems.

It is our conclusion that the role that the NRC plays in monitoring operator training contributes little and may actually aggravate the problem. NRC has a limited staff for supervising operator licensing, and many of these do not have actual experience in power plants. Therefore, NRC activities are limited to the administration of fairly routine licensing examinations and the spotchecking of requalification exams and training programs. In evaluating the training of operators to carry out emergency procedures, NRC failed to recognize basic faults in the procedures in existence at TMI. Since the utility has the tendency of equating the passing of an NRC examination with the satisfactory training of operators, NRC may be perpetuating a level of mediocrity.

The way that NRC evaluates the safety of proposed plants during the licensing process has a most unfortunate impact on the way operators are trained. Since during the licensing process applicants for licenses concentrate on the consequences of single failures, there is no attempt in the training program to prepare operators for accidents in which two systems fail independently of each other.

There were significant deficiencies in the management of the TMI-2 plant. Shift foremen were burdened with paper work not relevant to supervision and could not adequately fulfill their supervisory roles. There was no systematic check on the status of the plant and the line-up of valves when shifts changed. Surveillance procedures were not adequately supervised. And there were weaknesses in the program of quality assurance and control.

We agree that the utility that operates a nuclear power plant must be held legally responsible for the fundamental design and procedures that assure nuclear safety. However, the analysis of this particular accident raises the serious question of whether all electric utilities automatically have the necessary technical expertise and managerial capabilities for administering such a dangerous high-technology plant. We, therefore, recommend the development of higher standards of organization and management that a company must meet before it is granted a license to operate a nuclear power plant.

OVERVIEW

THE TRANSITION

We recognize that even with the most expeditious process for implementation, recommendations as sweeping as ours will take a significant amount of time to implement. Therefore, the Commission had to face the issue of what should be done in the interim with plants that are currently operating and those that are going through the licensing process.

The Commission unanimously voted:

> Because safety measures to afford better protection for the affected population can be drawn from the high standards for plant safety recommended in this report, the NRC or its successor should, on a case-by-case basis, before issuing a new construction permit or operating license: (a) assess the need to introduce new safety improvements recommended in this report, and in NRC and industry studies; (b) review, considering the recommendations set forth in this report, the competency of the prospective operating licensee to manage the plant and the adequacy of its training program for operating personnel; and (c) condition licensing upon review and approval of the state and local emergency plans.

A WARNING

During the time that our Commission conducted its investigation, a number of other reports appeared with recommendations for improved safety in nuclear power plants. While we are generally aware of the nature of these recommendations, we have not attempted a systematic analysis of them. Insofar as other agencies may have reached similar conclusions and proposed similar remedies, several groups arriving at the same conclusion should reinforce the weight of these conclusions.

But we have an overwhelming concern about some of the reports we have seen so far. While many of the proposed "fixes" seem totally appropriate, they do not come to grips with what we consider to be the basic problem. We have stated that fundamental changes must occur in organizations, procedures, and, above all, in the attitudes of people. No amount of technical "fixes" will cure this underlying problem. There have been many previous recommendations for greater safety for nuclear power plants, which have had limited impact. What we consider crucial is whether the proposed improvements are carried out by the same organizations (unchanged), with the same kinds of practices and the same attitudes that were prevalent prior to the accident. As long as proposed improvements are carried out in a "business as usual" atmosphere, the fundamental changes necessitated by the accident at Three Mile Island cannot be realized.

OVERVIEW

We believe that we have conscientiously carried out the mandate of the President of the United States, within our limits as human beings and within the limitations of the time allowed us. We have not found a magic formula that would guarantee that there will be no serious future nuclear accidents. Nor have we come up with a detailed blueprint for nuclear safety. And our recommendations will require great efforts by others to translate them into effective plans.

Nevertheless, we feel that our findings and recommendations are of vital importance for the future of nuclear power. We are convinced that, unless portions of the industry and its regulatory agency undergo fundamental changes, they will over time totally destroy public confidence and, hence, they will be responsible for the elimination of nuclear power as a viable source of energy.

A. Assessment of Significant Events

B. Health Effects

C. Public Health

D. Emergency Response

E. The Utility and Its Suppliers

F. Training of Operating Personnel

G. The Nuclear Regulatory Commission

H. The Public's Right to Information

COMMISSION FINDINGS

The President's Commission on the Accident at Three Mile Island, after conducting a study and investigation into the events of that accident and the conditions existing prior to the accident, finds and concludes*/:

<u>To prevent nuclear accidents as serious as Three Mile Island, fundamental changes will be necessary in the organization, procedures, and practices -- and above all -- in the attitudes of the Nuclear Regulatory Commission and, to the extent that the institutions we investigated are typical, of the nuclear industry.</u>

A. ASSESSMENT OF SIGNIFICANT EVENTS

   1. The accident at Three Mile Island (TMI) occurred as a result of a series of human, institutional, and mechanical failures.

   2. Equipment failures initiated the events of March 28 and contributed to the failure of operating personnel (operators, engineers, and supervisors) to recognize the actual conditions of the plant. Their training was deficient and left them unprepared for the events that took place. (See finding F.) These operating personnel made some improper decisions, took some improper actions, and failed to take some correct actions, causing what should have been a minor incident to develop into the TMI-2 accident.

---

*/ "Supplemental Views" from Commissioners are available and will be included in the permanent edition of the Commission's report.

COMMISSION FINDINGS

   3. The pilot-operated relief valve (PORV) at the top of the pressurizer opened as expected when pressure rose but failed to close when pressure decreased, thereby creating an opening in the primary coolant system -- a small-break loss-of-coolant accident (LOCA).*/ The PORV indicator light in the control room showed only that the signal had been sent to close the PORV rather than the fact that the PORV remained open. The operators, relying on the indicator light and believing that the PORV had closed, did not heed other indications and were unaware of the PORV failure; the LOCA continued for over 2 hours. The TMI-2 emergency procedure for a stuck-open PORV did not state that unless the PORV block valve was closed, a LOCA would exist. Prior to TMI, the NRC had paid insufficient attention to LOCAs of this size and the probability of their occurrence in licensing reviews. Instead, the NRC focused most of its attention on large-break LOCAs.

   4. The high pressure injection system (HPI) -- a major design safety system -- came on automatically. However, the operators were conditioned to maintain the specified water level in the pressurizer and were concerned that the plant was "going solid," that is, filled with water. Therefore, they cut back HPI from 1,000 gallons per minute to less than 100 gallons per minute. For extended periods on March 28, HPI was either not operating or operating at an insufficient rate. This led to much of the core being uncovered for extended periods on March 28 and resulted in severe damage to the core. If the HPI had not been throttled, core damage would have been prevented in spite of a stuck-open PORV.

   5. TMI management and engineering personnel also had difficulty in analyzing events. Even after supervisory personnel took charge, significant delays occurred before core damage was fully recognized, and stable cooling of the core was achieved.

   6. Some of the key TMI-2 operating and emergency procedures in use on March 28 were inadequate, including the procedures for a LOCA and for pressurizer operation. Deficiencies in these procedures could cause operator confusion or incorrect action.

---

*/ For a definition of loss-of-coolant accident and other technical terms used in the Commission's report, see the Glossary at the back of this volume.

COMMISSION FINDINGS

7. Several earlier warnings that operators needed clear instructions for dealing with events like those during the TMI accident had been disregarded by Babcock & Wilcox (B&W) and the Nuclear Regulatory Commission (NRC).

    a. In September 1977, an incident occured at the Davis-Besse plant, also equipped with a B&W reactor. During that incident, a PORV stuck open and pressurizer level increased, while pressure fell. Although there were no serious consequences of that incident, operators had improperly interfered with the HPI, apparently relying on rising pressurizer level. The Davis-Besse plant had been operating at only 9 percent power and the PORV block valve was closed approximately 20 minutes after the PORV stuck open. That incident was investigated by both B&W and the NRC, but no information calling attention to the correct operator actions was provided to utilities prior to the TMI accident. A B&W engineer had stated in an internal B&W memorandum written more than a year before the TMI accident that if the Davis-Besse event had occurred in a reactor operating at full power, "it is quite possible, perhaps probable, that core uncovery and possible fuel damage would have occurred."

    b. An NRC official in January 1978 pointed out the likelihood for erroneous operator action in a TMI-type incident. The NRC did not notify utilities prior to the accident.

    c. A Tennesse Valley Authority (TVA) engineer analyzed the problem of rising pressurizer level and falling pressure more than a year before the accident. His analysis was provided to B&W, NRC, and the Advisory Committee on Reactor Safeguards. Again no notification was given to utilities prior to the accident.

8. The control room was not adequately designed with the management of an accident in mind. (See also finding G.8.e.) For example:

    a. Burns and Roe, the TMI-2 architect-engineer, had never systematically evaluated control room design in the context of a serious accident to see how well it would serve in emergency conditions.

    b. The information was presented in a manner which could confuse operators:

        (i) Over 100 alarms went off in the early stages of the accident with no way of suppressing the unimportant ones and identifying the important ones. The danger of having too many alarms was recognized by Burns and Roe during the design stage, but the problem was never resolved.

        (ii) The arrangement of controls and indicators was not well thought out. Some key indicators relevant to the accident were on the back of the control panel.

COMMISSION FINDINGS

> (iii) Several instruments went off-scale during the course of the accident, depriving the operators of highly significant diagnostic information. These instruments were not designed to follow the course of an accident.
>
> (iv) The computer printer registering alarms was running more than 2-½ hours behind the events and at one point jammed, thereby losing valuable information.

c. After an April 1978 incident, a TMI-2 control room operator complained to his superiors about problems with the control room. No corrective action was taken by the utility.

9. In addition to the normal instrumentation present in the control room at the time of the accident, TMI-2 was equipped with a special data recorder that B&W had temporarily installed during the plant start-up and never removed. This data recorder, called a reactimeter, preserved a large amount of information useful in post-accident analysis. This type of data recorder was not required as standard equipment by the NRC.

10. Those managing the accident were unprepared for the significant amount of hydrogen generated during the accident. Indeed, during the TMI-2 licensing process which concentrated on large-break LOCAs, the utility represented and the NRC agreed that in the event of a large-break LOCA, the hydrogen concentration in containment would not be significant for a period of weeks. In the first 10 hours of the TMI accident (a small-break LOCA), enough hydrogen was produced in the core by a reaction between steam and the zirconium cladding and then released to containment to produce a burn or an explosion that caused pressure to increase by 28 pounds per square inch in the containment building. Thus, TMI illustrated a situation where NRC emphasis on large breaks did not cover the effects observed in a smaller accident.

11. Iodine filters in the auxiliary and fuel handling buildings did not perform as designed because the charcoal filtering capacity was apparently partially expended due to improper use before the accident. Required testing of filter effectiveness for the fuel handling building had been waived by the NRC. There were no testing requirements to verify auxiliary building filter effectiveness.

12. The nature and extent of damage to the core is not likely to be known with assurance until the core materials are recovered and carefully examined. However:

a. We estimate that there were failures in the cladding around 90 percent of the fuel rods. The interaction of the very hot cladding with water generated somewhere between 1,000 and 1,300 pounds of hydrogen gas and converted 44 to 63 percent of the zirconium to relatively weak zirconium oxide. As a result of oxidation and embrittlement of the fuel rod cladding, several feet

COMMISSION FINDINGS

of the upper part of the core fell into the gaps between the fuel rods, causing partial blocking of the flow of steam or water that could remove heat from the damaged fuel.

   b.   Fuel temperatures may have exceeded 4,000°F in the upper 30 to 40 percent of the core (approximately 30 to 40 tons of fuel). Temperatures in parts of the damaged fuel that were not effectively cooled by steam may have reached the melting point of the uranium oxide fuel, about 5,200°F.

   c.   An NRC study suggests that some of the fuel may have become liquid at temperatures above 3,500°F by dissolving in a zirconium-zirconium oxide mixture. The study estimates that the amount of fuel that may have melted by this process is from zero to a few tons. An independent analysis by Argonne National Laboratory suggests that the formation of such a mixture was unlikely.

   d.   Substantial fractions of the material in the reactor control rods melted.

   e.   There is no indication that any core material made contact with the steel pressure vessel at a temperature above the melting point of steel (2,800°F).

13.   The total release of radioactivity to the environment from March 28 through April 27 has been established as 13 to 17 curies of iodine and 2.4 million to 13 million curies of noble gases. (The health effects of the radiation released are described in finding B.)

   a.   Five hundred thousand times as much radioactive iodine (7.5 million curies) was retained in the primary loop. On April 1, 10.6 million curies of iodine were retained in the containment building's water and about 36,000 curies in the containment atmosphere. Four million curies were in the auxiliary building tanks. Almost all of the radioactive iodine released from the fuel was retained in the primary system, containment, and the auxiliary building. Since the accident, most of the short-lived radioactive iodine has decayed and is no longer a danger.

   b.   No detectable amounts of the long-lived radioactive cesium and strontium escaped to the environment, although considerable quantities of each escaped from the fuel to the water of the primary system, the containment building, and the auxiliary building tanks.

   c.   Most radioactivity escaping to the environment was in the form of fission gases transported through the coolant let-down/make-up system into the auxiliary building and through the building filters and the vent header to the outside atmosphere.

   d.   The major release of radioactivity on the morning of March 30 was caused by the controlled, planned venting of the make-up tank into the vent header. The header was known to have a leak.

COMMISSION FINDINGS

14. The process of recovery, cleanup, and waste disposal will be lengthy, costly, and presents its own health dangers. Cleanup of the reactor and auxiliary buildings and disposal of approximately one million gallons of radioactive water, a substantial amount of radioactive gases, and the solid radioactive debris within the reactor vessel remain to be done.

15. The cost of the accident, including this cleanup and a portion of the waste disposal, will be between $1 billion and $1.86 billion, if the plant can be refurbished. If it cannot be refurbished, the total cost will be significantly higher. An independent study prepared for the Commission estimates these costs as follows:

	Low	Medium	High
	(Millions of dollars)		
Refurbish TMI-2			
Emergency Management	$ 120	$ 160	$ 225
Replacement Power*/	678	966	1,128
Plant Refurbishment	249	306	503
Total**/	$1,047	$1,432	$1,856

16. The 1974 WASH 1400 Reactor Safety Study (the Rasmussen Report) analyzed events, equipment failures, and human errors that could happen during reactor accidents, including those associated with the TMI accident. However, NRC has not made systematic use of WASH 1400, a major study commissioned by the Atomic Energy Commission (AEC), in its design review analyses. WASH 1400 showed that small-break LOCAs similar in size to the accident at TMI were much more likely to occur than the design basis large-break LOCAs, and can lead to the same consequences. Further, the probability of occurrence of an accident like that at Three Mile Island was high enough, based on WASH 1400, that since there had been more than 400 reactor years of nuclear power plant operation in the United States, such an accident should have been expected during that period.

---

*/ The low case assumes TMI-2 will be returned to service in January 1983, the medium assumes January 1984, and the high assumes January 1985.

**/ The costs associated with health effects have been deleted from this table. The costs projected by the study had a minimal effect on the total costs projected. The Commission believes that the analysis of health effects costs was insufficient to reach the conclusion set out in the study.

COMMISSION FINDINGS

17. The Commission tried to determine what would have happened if certain additional events had occurred during the accident. For a discussion of these scenarios, see the Commission Overview and the technical staff analysis report on "Alternative Event Sequences."

COMMISSION FINDINGS

B. HEALTH EFFECTS

1. Based on available dosimetric and demographic information:

a. It is estimated that between March 28 and April 15, the collective dose resulting from the radioactivity released to the population living within a 50-mile radius of the plant was approximately 2,000 person-rems. The estimated annual collective dose to this population from natural background radiation is about 240,000 person-rems. Thus, the increment of radiation dose to persons living within a 50-mile radius due to the accident was somewhat less than one percent of the annual background level. The average dose to a person living within 5 miles of the nuclear plant was calculated to be about 10 percent of annual background radiation and probably was less.

b. The maximum estimated radiation dose received by any one individual in the off-site general population (excluding the plant workers) during the accident was 70 millirems. On the basis of present scientific knowledge, the radiation doses received by the general population as a result of exposure to the radioactivity released during the accident were so small that there will be no detectable additional cases of cancer, developmental abnormalities, or genetic ill-health as a consequence of the accident at TMI.

c. During the period from March 28 to June 30, three TMI workers received radiation doses of about 3 to 4 rems; these levels exceeded the NRC maximum permissible quarterly dose of 3 rems.

d. The process of recovery and cleanup presents additional sources of possible radiation exposure to the workers and the general population.

2. There were deficiencies in instrumentation for measuring the radioactivity released, particularly during the early stages of the accident. However, these deficiencies did not affect the Commission staff's ability to estimate the radiation doses or health effects resulting from the accident.

COMMISSION FINDINGS

3. The health effects of radiation dose levels of a few rems or less are not known. Estimates of the potential health effects of the TMI accident are based on extrapolations from the known health effects of higher levels of radiation.

4. The major health effect of the accident appears to have been on the mental health of the people living in the region of Three Mile Island and of the workers at TMI. There was immediate, short-lived mental distress produced by the accident among certain groups of the general population living within 20 miles of TMI. The highest levels of distress were found among adults a) living within 5 miles of TMI, or b) with preschool children; and among teenagers a) living within 5 miles of TMI, b) with preschool siblings, or c) whose families left the area. Workers at TMI experienced more distress than workers at another plant studied for comparison purposes. This distress was higher among the nonsupervisory employees and continued in the months following the accident.

COMMISSION FINDINGS

C. PUBLIC HEALTH

1. The Nuclear Regulatory Commission has primary responsibility and regulatory authority for health and safety measures as they relate to the operation of commercial nuclear plants. While the NRC has certain requirements in connection with radiation exposure and medical monitoring of workers at nuclear plants, it has no requirements for medical examination of workers other than licensed reactor operators, and even those examinations are only performed to assure that the operators do not have physical or mental conditions that might impair their ability to perform their jobs safely. Metropolitan Edison's (Met Ed) administrative procedures go beyond this NRC requirement and provide that all radiation workers receive routine medical examinations to assess any possible radiation-related illnesses. The NRC only requires monitoring and reporting of radiation exposure for workers who, in the utility's view, are likely to receive doses beyond NRC-specified levels. Met Ed does not keep, and the NRC does not require it to report, a record of the total radiation exposure of workers from both occupational and nonoccupational (for example, medical and dental) sources.

2. The Public Health Service agencies of the U.S. Department of Health, Education, and Welfare (HEW),*/ whose sole mission is protection and promotion of the public health, have very limited responsibilities with respect to radiological health matters relating to the location, construction, and routine operation of commercial nuclear power plants.

3. Although there were designated channels of communication and specific responsibilities assigned for federal agencies responding to the radiological emergency at TMI (for example, Interagency Radiological Assistance Plan), the existence of these channels and responsibilities was generally unknown to many high-level federal officials. In several instances during the course of the accident, some federal agencies were unaware of what other federal agencies were doing in providing support personnel and resources.

---

*/   Now the U.S. Department of Health and Human Services.

COMMISSION FINDINGS

4. Research on the biological effects of ionizing radiation is conducted and/or sponsored by a number of federal agencies. In fiscal year 1978, the federal government spent approximately $76.5 million on such research. More than 60 percent of this funding was provided by the U.S. Department of Energy. With the exception of potassium iodide, there are no drugs presently approved by the Food and Drug Administration for the prevention or mitigation of adverse effects of ionizing radiation.

5. States have primary responsibility for protecting the health and safety of their citizens. Pennsylvania public health officials and health-care providers in the TMI area did not have sufficient resources to respond to the potentially serious health consequences of the accident at TMI. Responsibility for radiological protection in Pennsylvania rests with the Department of Environmental Resources (DER). At the time of the accident, the Pennsylvania Department of Health was not organized to respond to radiological emergencies, and maintained no formal liaison with DER on radiological health matters.

6. During the accident, TMI-area hospital administrators found no one at the state level with authority to recommend when to evacuate patients and when to resume normal admitting procedures. The Pennsylvania Secretary of Health viewed his department's role with respect to area hospitals as informational, not advisory.

7. During the first days of the accident, Met Ed did not notify its physicians under contract who would have been responsible for the on-site treatment of injured, contaminated workers during the accident. The emergency radiological medical care training provided to these physicians to provide on-site emergency care to such workers was inadequate.

8. Met Ed experienced several radiation protection problems during the accident: a) the emergency control center for health physics operations and the analytical laboratory to be used in emergencies was located in an area that became uninhabitable in the early hours of the accident; b) there was a shortage of respirators; and c) there was an inadequate supply of uncontaminated air.

9. NRC regulations on health physics education of nuclear power plant workers leaves the details of such things as course content, frequency, and attendance to the discretion of the licensee, subject to NRC inspection. Similarly, NRC regulations for environmental radiological monitoring leaves the details and methods of how these requirements are to be implemented (for example, types of dosimeters, kind of sample analysis) to the discretion of the licensee, subject to NRC inspection and approval.

COMMISSION FINDINGS

D. EMERGENCY RESPONSE

1. Planning for the protection of the public in the event of a radiological release that extends beyond the boundary of TMI was highly complex. It involved the utility and government agencies at the local, state, and federal levels. That complexity posed problems in the case of the accident at Three Mile Island; some of the written plans that existed had not been coordinated and contained different systems for classifying accidents and different guidelines for notifying government officials.

2. In approving sites for reactors, the NRC has required licensees to plan for off-site consequences of radioactive releases only within the "low population zone" (LPZ), an area containing "residents, the total number and density of which are such that there is a reasonable probability that appropriate protective measures could be taken in their behalf in the event of a serious accident." As calculated for the design-basis accident for TMI-2, this zone was a 2-mile radius.

3. Emergency planning had a low priority in the NRC and the AEC before it. There is evidence that the reasons for this included their confidence in designed reactor safeguards and their desire to avoid raising public concern about the safety of nuclear power.

4. The NRC has not made the existence of a state emergency or evacuation plan a condition for plant licensing. A state may voluntarily submit a response plan to NRC for concurrence, and if the plan meets NRC guidelines -- which do not have the force of law -- the state receives a formal letter of concurrence. At the time of the accident, Pennsylvania did not have an NRC concurred-in plan. The NRC concurrence program has been called ineffective by federal and state emergency preparedness officials.

5. The utility has the responsibility to prevent or to mitigate off-site radiation releases and to notify the government agencies designated in its emergency plan in the event that an emergency is declared. Federal, state, and local agencies are responsible for off-site response to radiation releases. At the

COMMISSION FINDINGS

time of the TMI accident, the local and county governments had the primary action role once notified of the emergency. None of the local communities within the 5-mile radius of the plant had emergency plans, and the existing county plans did not include detailed evacuation plans.

6. At all levels of government, planning for the off-site consequences of radiological emergencies at nuclear power plants has been characterized by a lack of coordination and urgency. For example, a federal response plan in preparation since 1974 by federal emergency preparedness agencies was unfinished at the time of the accident because of an interagency jurisdictional dispute and lack of communication. Pennsylvania did not begin to develop a radiological emergency plan until 1975, even though nuclear power plants had been operating within its borders for at least a year prior to that time. People who attempted to generate interest in radiological emergency planning at the local level near TMI found local officials apathetic. Part of the reason for this was the attitude fostered by the NRC regulatory approach, and by Met Ed at the local level, that radiological accidents having off-site consequences beyond the LPZ were so unlikely as not to be of serious concern.

7. Interaction among NRC, Met Ed, and state and local emergency organizations in the development, review, and drill of emergency plans was insufficient to ensure an adequate level of preparedness for a serious radiological incident at TMI.

8. Although NRC personnel were on-site within hours of the declaration of a site emergency and were in constant contact with the utility, the NRC was not able to determine and to understand the true seriousness and nature of the accident for about 2 days, when the fact of extensive core damage and the existence of the hydrogen bubble were generally recognized within NRC.

9. During the first 2-½ days of the accident, communications between the NRC Incident Response Center in Bethesda, Maryland, where the senior management was located, and the site were such that senior management officials found it extremely difficult to obtain up-to-date information. Communications were so poor on Friday morning that the senior management could not and did not develop a clear understanding of conditions at the site. As a result, an evacuation was recommended to the state by the NRC senior staff on the basis of fragmentary and partially erroneous information. Communications did not improve until Harold Denton, designated the sole source of information, arrived on the site and communicated with NRC headquarters, the Governor's office, and the White House by White House communications line.

10. The reality of possible evacuation was quite different from the theoretical planning requirements imposed by the NRC and Pennsylvania before the accident. The 5-mile emergency plans were developed according to a Pennsylvania requirement for emergency planning within a 5-mile radius of nuclear power plants. The Pennsylvania requirement was stricter than that prescribed by NRC,

COMMISSION FINDINGS

which only required TMI to have a plan for a radius to 2 miles. (See finding D.2.) It is known that the consequences of a postulated major release to the atmosphere from a reactor accident could lead to significant doses of radiation being received many tens of miles from the site of the accident. At TMI-2, although the radiation releases were significantly lower than the design-basis accident, evacuation was being considered for distances much greater than 2 miles. During the TMI accident, NRC believed that the consequences of the accident might extend far beyond the 2- or 5-mile radius. As a result, evacuation plans were hurriedly developed for distances of 10 and 20 miles from the plant.

11. During the most critical phase of the accident, the NRC was working under extreme pressure in an atmosphere of uncertainty. The NRC staff was confronted with problems it had never analyzed before and for which it had no immediate solutions. One result of these conditions was the calculational errors concerning the hydrogen bubble, which caused the NRC to misunderstand the true conditions in the reactor for nearly 3 days.

12. On Friday and Saturday, certain NRC officials incorrectly concluded that a hydrogen bubble in the reactor vessel would soon contain enough oxygen to burn or explode. Ignoring correct information supplied by a B&W employee and certain members of its own staff, NRC relied instead upon incorrect information supplied by other members of its staff and by others that sufficient oxygen was being formed from water radiolysis to reach a concentration sufficient for a burn or explosion. Based on this information, the NRC commissioners began formulating new recommendations for evacuation. On Sunday, NRC staffers obtained information from several national laboratories and from General Electric and Westinghouse that sufficient oxygen could not form. The Sunday information ended the concern about oxygen formation and evacuation. This incident suggests that NRC lacks sufficient knowledge and expertise in water radiolysis.

13. The role of the NRC commissioners and their decision-making process during the accident were ill-defined. Although the commissioners on Friday assumed responsibility for making recommendations to the Governor concerning protective action, there was no apparent procedure by which issues and staff recommendations were explored and resolved. The commissioners were preoccupied with matters such as the details of evacuation planning and the drafting of a press release.

14. Existing emergency plans were not designed to meet the demands of a protracted crisis. The plans had no mechanisms for establishing reliable communications among the on-site and the several off-site organizations responsible for various aspects of the emergency response.

15. There were no hospitals within 5 miles of TMI, but there were several hospitals within the expanded, proposed evacuation zones. The NRC estimated that it would be able to give officials a few hours "lead time" for evacuation. But hospital administrators

COMMISSION FINDINGS

estimated they would need substantially more time to evacuate patients.

16. During the TMI accident, the actual radiation levels outside the plant were low, but there was uncertainty about the possibility of serious releases on short notice. Federal and state officials disagreed about the nature of the information on which to base evacuation decisions and other protective actions during the emergency. Some officials based their decisions on actual radiation exposure levels, while others based their decisions on concerns about potential releases of large amounts of radioactivity. For example, the Pennsylvania Bureau of Radiation Protection told the Governor on Friday that radiation levels indicated that no protective action of any kind was required; on that same morning, NRC Chairman Hendrie recommended that pregnant women and young children be advised to leave the area near the plant because of his concern about possible releases of radioactivity.

17. At approximately 12:30 p.m., March 30, Governor Thornburgh advised pregnant women and preschool aged children to leave the area within a 5-mile radius of TMI until further notice. A substantial number of other persons, including health professionals, voluntarily left the area around the plant during the weekend of March 30 through April 1. The advisory to pregnant women and preschool children was lifted on April 9.

18. Throughout the accident, the Pennsylvania Emergency Management Agency (PEMA) received reports concerning conditions at the site from the Bureau of Radiation Protection. During the first 2 days of the accident, however, the director of PEMA also received background information on the status of the plant from the Governor's office by attending meetings and press conferences and relayed that information to county organizations, which, in turn, informed the local civil defense directors. Starting Saturday, the PEMA director was no longer included in these meetings with the result that PEMA and county and local civil defense organizations had to rely primarily on the news media for information about conditions at the site. They found this an unsatisfactory source of information and believed that this arrangement compromised their effectiveness in responding to the accident.

19. The TMI emergency plan did not require the utility to notify state or local health authorities in the event of a radiological accident. (See also finding C.7.)

20. For over 25 years, the use of blocking agents such as potassium iodide to prevent the accumulation of radioiodine in the thyroid gland has been known. The effectiveness of potassium iodide administration for thyroid gland protection in the event of releases of radioiodine was recognized by the National Council on Radiation Protection and Measurement in 1977. The Food and Drug Administration authorized use of potassium iodide as a thyroid-blocking agent for the general public in December 1978. However, at the time of the TMI accident, potassium iodide for this use was not commercially available in the United States in

COMMISSION FINDINGS

quantities sufficient for the population within a 20-mile radius of TMI. At the time of the accident, Met Ed had no supply of potassium iodide on-site. A crash effort by the federal government and private industry resulted in delivery of substantial supplies of potassium iodide to Pennsylvania within 2 days of the decision to obtain such supplies.

COMMISSION FINDINGS

E. THE UTILITY AND ITS SUPPLIERS

    1. In a number of important cases, General Public Utilities Corporation (GPU), Met Ed, and B&W failed to acquire enough information about safety problems, failed to analyze adequately what information they did acquire, or failed to act on that information. Thus, there was a serious lack of communication about several critical safety matters within and among the companies involved in the building and operation of the TMI-2 plant. A similar problem existed in the NRC. (See finding G.)

    a. The September 1977 incident at Davis-Besse, another plant with a B&W reactor, foreshadowed several aspects of the TMI-2 accident. A serious warning by a senior engineer at B&W that more precise instructions be given to operators "fell between the cracks." This warning, issued 13 months before the TMI-2 accident, if heeded, could have prevented the accident. (See also finding A.7.a.)

    b. Nine times before the TMI accident, PORVs stuck open at B&W plants. B&W did not inform its customers of these failures, nor did it highlight them in its own training program so that operators would be aware that such a failure causes a small-break LOCA.

    c. A report by an engineer at TVA questioning how operators might respond to rising pressurizer level and falling pressure was sent to B&W in April 1978. B&W took 9 months to respond and never advised its utility customers of the concern expressed in the report. The concern was similar to the one which B&W itself had identified from the Davis-Besse incident.

    d. TMI-2 had repeated problems with the condensate polishers. During the 18-month period before the accident, no effective steps were taken to correct these problems. These polishers probably initiated the March 28 sequence of events.

    e. The TMI-2 operators had never had specific training about the dangers of saturation conditions in the core, although they were generally familiar with the concept. Although Met Ed

COMMISSION FINDINGS

believed saturation had occurred in an incident a year before the accident that could have led to core uncovery, its hazards were not emphasized to the operators. When saturation occurred again on March 28, operators did not recognize the significance of that fact and take corrective action promptly.

    f. After an incident at TMI-2 a year earlier during which the PORV stuck open, an indicator light was installed in the control room. That light showed only that a signal had been sent to close the valve -- it did not show whether the valve was actually closed -- and this contributed to the confusion during the accident. (See finding A.3.)

Timely attention to all of these factors probably would have prevented the accident.

    2. The GPU Service Corporation (GPUSC) had final responsibility for design of the plant. However, by its own account, it lacked the staff or expertise in certain areas to discharge that responsibility. Once construction was complete, GPUSC turned the plant over to Met Ed to run, but Met Ed did not have sufficient knowledge, expertise, and personnel to operate the plant or maintain it adequately.

    3. Responsibility for management decisions was divided among the TMI site, Met Ed, and GPU. GPU recognized in early 1977 that integration of operating responsibility into one organization was desirable. A management audit by Booz, Allen, and Hamilton completed in the spring of 1977 recommended clarifying and reevaluating the roles of GPUSC and Met Ed in the design and construction of new facilities; strengthening communications between GPUSC and Met Ed; and establishing minimum standards for the safe operation of GPU's nuclear plants. However, integration of management did not occur until after the accident.

    4. The Met Ed management systems, procedures, and practices did not provide Met Ed a firm understanding of TMI's operations, nor were effective systems of checks and balances in use.

    a. Met Ed had a plan for a quality assurance program that met NRC requirements. The NRC requirements, however, were inadequate because they did not require quality assurance programs to be applied to the plant as a whole, but rather only to systems classified as "safety-related." Neither the PORV nor the condensate polishers were classified as "safety-related." In addition, the NRC did not require the level of independent review (i.e., outside of line management) normally found in the quality assurance programs of safety-critical industries.

    b. There was no requirement for an independent (i.e., outside of line management) safety assessment of operating procedures. Independent audit of the performance of surveillance procedures was required only every 2 years.

COMMISSION FINDINGS

c. Met Ed's implementation of its own quality assurance plan was found to contain significant deficiencies by the Commission staff and in an NRC post-accident audit of TMI-2. For example:

(i) There were not enough inspectors to do the inspections required under the Met Ed plan.

(ii) The NRC audit reported deficiencies in maintaining "as built" drawings and in the purchasing of "safety-related" equipment without quality controls.

(iii) Although all plant procedures were required to be reviewed every 2 years, there was no plan for such a review and no review had in fact been made of those TMI-2 procedures that were more than 2 years old.

(iv) Although such inspections were required, Met Ed had not scheduled or conducted any inspections of materials, components, or equipment in storage.

(v) There were deficiencies in the reporting, analysis, and resolution of problems in "safety-related" equipment and other events required to be reported to the NRC.

(vi) Independent assessment of general plant operations was minimal.

d. Met Ed did not go beyond NRC requirements in such areas as:

(i) Requiring reporting, resolution, and trending of problems in plant equipment and procedures which were not "safety-related."

(ii) Applying its quality assurance program to the operation of non-"safety-related" equipment and systems vital to plant operation, consistent with the importance of those systems to safety. For example, no quality assurance review was given to radiation monitoring equipment, control rod drive mechanisms, hydrogen recombiners, the PORV, or condensate polishers. In addition, Met Ed's quality assurance program was not applied to the maintenance or the procedures associated with such non-"safety-related" equipment.

As a result of these deficiencies, the safe operation of the TMI-2 plant was impaired.

5. Utility management did not require attention to detail as a way of life at Three Mile Island. For example:

COMMISSION FINDINGS

        a. Management permitted operation of the plant with a number of poor control room practices:

        (i) A shift supervisor testified that there had never been less than 52 alarms lit in the control room.

        (ii) TMI Commission staff and NRC inspections noted a large number of control room instruments out of calibration and tags hanging on the instrument panel indicating equipment out of service. Operators testified that one of these tags obscured one of the emergency feedwater block control valve indicator lights.

        (iii) When shifts changed in the control room, there was no systematic check on the status of the plant and the line-up of valves.

        b. There were deficiencies in the review, approval, and implementation of TMI-2 plant procedures.

        (i) Although Met Ed procedures required closing the PORV block valve when temperatures in the tailpipe exceeded 130°F, the block valve had not been closed at the time of the accident even though temperatures had been well above 130°F in the tailpipe for weeks.

        (ii) Operators were not given adequate information about temperatures to be expected in the PORV tailpipe after the PORV opened.

        (iii) A 1978 B&W analysis of a certain kind of small-break LOCA was misinterpreted by Met Ed. That misinterpretation was incorporated by Met Ed into the LOCA emergency procedure available at the time of the accident.

        (iv) Operating and emergency procedures that had been approved by Met Ed and were in use at the time of the accident contained many minor substantive errors, typographical errors, and imprecise or sloppy terminology. Some were inadequate. (See finding A.6.)

        (v) A 1978 revision in the TMI-2 surveillance procedure for the emergency feedwater block valves violated TMI-2's technical specifications, but no one realized it at the time. The approval of the revision in the surveillance procedure was not done according to Met Ed's own administrative procedures.

46

COMMISSION FINDINGS

> (vi) Performance of surveillance tests was not adequately verified to be sure that the procedures were followed correctly. On the day of the accident, emergency feedwater block valves which should have been open were closed. They may have been left closed during a surveillance test 2 days earlier.

    c. There were deficiencies in maintenance:

> (i) After the accident, valves in the TMI-1 containment building exhibited long-term lack of maintenance. Boron stalactites more than a foot long hung from the valves and stalagmites had built up from the floor.
>
> (ii) Review of equipment history for the 6 months prior to the accident showed that a number of equipment items that figured in the accident had had a poor maintenance history without adequate corrective action. These included the pressurizer level transmitter, the hydrogen recombiner, pressurizer heaters, make-up pump switches, and the condensate polishers.
>
> (iii) Despite a history of problems with the condensate polishers, procedures were not changed to ensure that operators would bypass the polishers during maintenance operations to protect the plant from a possible malfunction of the polisher.

    d. After the accident, radiological control practices were observed to be deficient. Contaminated and potentially contaminated equipment was found in uncontrolled areas of the auxiliary building.

    e. Training of operators and supervisors did not give sufficient emphasis to a fundamental understanding of the reactor. There was no comprehensive evaluation of operator performance on the job to meet the requalification requirements of 10 CFR 55. (See finding F.)

    f. Reports of operating experience at other plants were screened by technical analysts who did not have nuclear backgrounds. They were given no instruction in how to screen such operating reports, according to Station Manager Gary Miller. The technical analysts routed experience summaries to designated people at TMI. The routing sometimes took several months. The person in the training department who was assigned to review these summaries often did not get to them for several months after he received them because of the press of other work. The training department held only one 2-hour class per year on operating experience at other plants.

COMMISSION FINDINGS

g. There was no group with special responsibility for receiving and acting upon potential safety concerns raised by employees.

h. Management did not assure adequate identification of piping and valves throughout the plant. The Commission staff noted that pipe and valve identification practices were significantly below standard industrial practices. Eight hours into the accident, Met Ed personnel spent 10 minutes trying unsuccessfully to locate three decay heat valves in a high radiation field in the auxiliary building.

i. Management did not assure that Licensee Event Reports (LER) met basic NRC requirements. A review of TMI-2's LERs disclosed repeated omissions, inadequate failure analyses, and inadequate corrective actions.

j. Met Ed did not correct deficiencies in radiation monitoring equipment, although the deficiencies were pointed out by an NRC audit months before the accident.

k. On November 3, 1978, a mechanic caused a complete shutdown of the plant, including exercising of emergency systems, when he tripped a switch on the polisher electrical panel, thinking he was turning on a light. The only corrective action was to put a guard on the switch.

l. Sensitive areas of the plant were accessible to large numbers of people. On the day before the accident, as many as 750 people had access to the auxiliary building.

m. The manual control station of the polisher bypass valve was nearly inaccessible and took great effort, in a physically awkward position, to operate.

n. Iodine filters were left in continuous use rather than being preserved to filter air in the event of radioactive contamination. As a result, they did not have full capacity on the day of the accident. (See finding A.11.).

COMMISSION FINDINGS

F. TRAINING OF OPERATING PERSONNEL

1. Training of Met Ed operators and supervisors was inadequate and contributed significantly to the seriousness of the accident. The training program gave insufficient emphasis to principles of reactor safety.

2. The TMI training program conformed to the NRC standard for training. Moreover, TMI operator license candidates had higher scores than the national average on NRC licensing examinations and operating tests. Nevertheless, the training of the operators proved to be inadequate for responding to the accident.

3. NRC standards allowed a shallow level of operator training.

   a. The Operator Licensing Branch activities were principally restricted to preparing and giving initial licensing examinations and occasional visits to vendors for an informal spot check of start-up certification tests. The branch was heavily involved in the initial start-up of the B&W cold licensing program in the early 1970s. A paper review of B&W's course for new plant operator training was performed without comment in 1976.

   b. NRC prescribed only minimal requirements for operator training. There were no minimum educational requirements for operators; there was no requirement for checks to be made on the psychological fitness of candidates or whether they had criminal records.

   c. An individual could fail parts of either the NRC licensing examination or the utility requalification examination, including sections on emergency procedures and equipment, and still pass the overall examination by getting a passing average score, and qualify to operate the reactor.

   d. The NRC had no criteria for the qualifications of those individuals who carry out the operator training program. It also did not conduct regular in-depth reviews of the training programs.

COMMISSION FINDINGS

4. Met Ed had primary responsibility for the training of operators. The quality of the training program at TMI was low.

    a. The training program was quantitatively and qualitatively understaffed as well as conceptually weak; emphasis was not given to fundamental understanding of the reactor and little time was devoted to instruction in the biological hazards of radiation. The content was left to the instructors, who had no greater formal educational qualifications than those of their students.

    b. TMI-2's station manager, unit superintendent, and supervisor of operations were not involved in operator training.

    c. With NRC approval, the unit superintendent and the station manager at TMI were only required to acquire the experience and training necessary to be examined for a senior reactor operator license, but were not required to hold such a license.

    d. Although auxiliary operators performed tasks that could affect reactor power level or involve the handling of radioactive material, there was no formally defined training program for them.

    e. Met Ed did not request waivers from employees with naval reactor experience to allow examination of their Navy records.

5. TMI contracted with B&W to carry out a portion of the TMI operator training. B&W performed only those functions specifically required under the agreement.

    a. There was little interaction in B&W between those who conducted training and those responsible for nuclear plant design. Course content and conduct of courses were made up by the B&W training department, entirely on its own. There were no formal syllabi or training manuals.

    b. The simulator at B&W was a key tool in the training of operators. Simulator training did not include preparation of the operators for multiple-failure accidents. Indeed, the B&W simulator was not, prior to March 28, programmed to reproduce the conditions that confronted the operators during the accident. It was unable to simulate increasing pressurizer level at the same time that reactor coolant pressure was dropping.

COMMISSION FINDINGS

G. THE NUCLEAR REGULATORY COMMISSION

1. A purpose of the Energy Reorganization Act of 1974 was to divorce the newly created NRC from promotion of nuclear power. According to one of the present NRC commissioners, "I still think it [the NRC] is fundamentally geared to trying to nurture a growing industry." We find that the NRC is so preoccupied with the licensing of plants that it has not given primary consideration to overall safety issues.

2. NRC labels safety problems that apply to a number of plants as "generic." Once a problem is labeled "generic," the licensing of an individual plant can be completed without resolving the problem. NRC has a history of leaving generic safety problems unresolved for periods of many years -- for example, the problem of anticipated transients without scram. In 1976 during the TMI-2 operating license (OL) review, the Advisory Committee on Reactor Safeguards recommended, as they did in at least one other OL review, that prior to commercial operation further evaluations be done of various possible accidents, including low-probability accidents. NRC staff designated this as a "generic issue." TMI-2 received its OL 2 years later without the resolution of the issue.

3. Although NRC accumulates an enormous amount of information on the operating experience of plants, there was no systematic method of evaluating these experiences and looking for danger signals of possible generic safety problems. In 1978, the General Accounting Office criticized NRC for this failure, but no corrective action had been taken as of the TMI-2 accident.

4. The NRC commissioners have largely isolated themselves from the licensing process. Although the commissioners have adopted unnecessarily stringent _ex parte_ rules to preserve their adjudicative impartiality, they have also delegated most of their adjudicative duties to the Atomic Safety and Licensing Appeal Board and actually adjudicate approximately 25 percent of all licensing decisions. That figure is misleadingly high, in part because a number of the decisions do not represent significant adjudicatory events and include decisions on exports. The commissioners have also isolated themselves from the overall management of the NRC.

COMMISSION FINDINGS

One of the present NRC commissioners, testifying before Congress, said, "There has, I think, been too little Commission involvement in the setting of safety policy in this agency and too little Commission guidance on safety matters to the staff and to the board."

5. The major offices within the NRC operate independently with little evidence of exchange of information or experience. For example, the fact that operators could be confused due to reliance on pressurizer level had been raised at various levels within the NRC organization. Yet, the matter "fell between the cracks" and never worked its way out of the system prior to the TMI-2 accident.

6. Licensing of a nuclear plant is a two-step process. First, the company must obtain a construction permit (CP) and several years later must obtain an operating license (OL). The CP stage does not require complete design plans, and therefore the full safety review does not occur until the OL stage. By then, hundreds of millions of dollars have been spent or committed in the construction process. Therefore, the ultimate safety review may be influenced by economic considerations that can lead to a reluctance to order major changes at the OL stage.

7. The Advisory Committee on Reactor Safeguards (ACRS) reviews all applications for licenses and poses whatever questions it deems appropriate. The ACRS is the only body independent of the NRC staff which regularly reviews safety questions. However, it has established no firm guidelines or procedures, and generally has only monthly meetings of limited duration. ACRS members are part-time and have a very small staff, thus they must rely heavily on the NRC staff for follow-up of their concerns. ACRS members tend to concentrate on their own particular areas of expertise, thereby resulting in a fragmented licensing review.

8. There are serious inadequacies in the NRC licensing process.

    a. Applicants for licenses are only required to analyze "single-failure" accidents; they are not required to analyze what happens when two systems or components fail independently of each other. The accident at TMI-2 was a multiple-failure accident.

    b. NRC's design safety review places primary emphasis on those items labeled "safety-related." This designation is crucial since items not labeled "safety-related" need not be reviewed in the licensing process, are not required to meet NRC design criteria, need not be testable, do not require redundancy, and are ordinarily not subject to NRC inspection. There are no precise criteria as to which components and systems are to be labeled "safety-related;" the utility makes the initial determination subject to NRC approval. For example, at TMI-2, the PORV was not a "safety-related" item because it had a block valve behind it. On the other hand, the block valve was not "safety-related" because it had a PORV in front of it.

COMMISSION FINDINGS

      c. NRC's reliance upon artificial categories of "safety-related" items has caused it to miss important safety issues and has led the nuclear industry to merely comply with NRC regulations and to equate that compliance with operational safety. Thus, over-emphasis by the NRC process on specific categories of items labeled "safety-related" appears to interfere with the development, throughout the nuclear industry, of a comprehensive safety consciousness, that is, a dynamic day-to-day process for operating safely.

      d. There is no identifiable office within NRC responsibile for systems engineering examination of overall plant design and performance, including interactions between major systems.

      e. There is no office within NRC that specifically examines the interface between machines and human beings. There seems to be a persistent assumption that plant safety is assured by engineered equipment, and a concomitant neglect of the human beings who could defeat it if they do not have adequate training, operating procedures, information about plant conditions, and manageable monitors and controls. For example, despite recognition within NRC and various industrial groups that outdated technology in the control room could seriously handicap operators during an accident, NRC continues to license new plants with similarly deficient control rooms. As noted before, problems with the control room contributed to the confusion during the TMI accident. (See also finding A.8.)

      f. The requirement of additional instrumentation to aid in accident diagnosis and control was considered by NRC as early as 1975, but its implementation was delayed by industry opposition as expressed by the Atomic Industrial Forum (AIF). AIF opposition was based on, among other things, the belief that the instrumentation required would cover "Class 9" accidents, and therefore, would extend beyond design-basis requirements. The lack of instrumentation to display in the control room the full range of temperatures from the core thermocouples contributed to the confusion involved in the attempt to rapidly depressurize the primary system on March 28.

      g. It is common to issue operating licenses to plants when there are still "open safety items." When a plant is licensed with many open items, the Division of Operating Reactors, which has the technical expertise to supervise operating plants, may refuse to accept jurisdiction from the Division of Project Management. In effect, the plant then ends up in a regulatory "limbo," receiving insufficient attention from either division. TMI-2 was in this "limbo" at the time of the accident, 13 months after its OL had been issued.

      h. When NRC issues new guidelines concerning safety, there is usually no systematic review, on a plant-by-plant basis, of operating plants and plants under construction for possible "backfitting." For example, Chairman Hendrie explained to a Congressional committee that stricter requirements for on- and

COMMISSION FINDINGS

off-site emergency plans had not been imposed on any already operating plants because of the need to balance costs against safety. The committee, however, found no significant cost burden in requiring utilities to upgrade and implement emergency plans. Similarly, NRC determined not to backfit the 1975 Standard Review Plan (SRP) to those plants, such as TMI-2, that received construction permits prior to September 1, 1975. According to Roger Mattson, director of the Division of Systems Safety, if individual SRP requirements had been reviewed for possible backfitting, the SRP requirement of diverse containment isolation actuation would probably have been backfitted to plants such as TMI-2. Instead, TMI-2 containment was isolated only when the pressure in the building exceeded 4 pounds per square inch. Thus, containment isolation did not occur until several hours after the start of the accident. However, this delay had little effect on the actual small releases of radioactive material during the accident.

    i. Although decisions of significant public health impact are considered in the licensing process, NRC has no specific mechanism for interactions with public health agencies in the licensing process, other than the U.S. Environmental Protection Agency (EPA) which does review Environmental Impact Statements filed by applicants for CPs and OLs.

    9. The Office of Inspection and Enforcement (I&E) is charged with determining whether licensees are complying with NRC regulations, rules, and licensing conditions. Some serious deficiencies in this office are:

    a. A 1978 General Accounting Office report found that I&E inspectors did little independent testing of construction work, relied heavily on the utility's self-evaluation, spent little time observing ongoing construction work, and did not communicate routinely with people who did the actual construction work. Similar problems exist in I&E inspections of operating plants. For example, the principal I&E inspector for TMI-2 completed an inspection shortly before the accident by examining utility records and interviewing plant personnel, but without physically examining any equipment.

    b. A 1978 survey of I&E commissioned by the NRC determined that the majority of inspectors felt their procedures were unclear and lacking in sufficient technical guidance.

    c. Of crucial significance to I&E's system of inspection and enforcement are the Licensee Event Reports (LER) in which utilities report and evaluate important incidents. However, both licensees and vendors often have a strong financial disincentive to evaluate and report safety problems that may result in more stringent regulations, at least in part because it is uncertain which entity will ultimately bear the cost of increased safety. I&E makes little effort to systematically review the LERs, has no formal review mechanism for them, and hence, must rely on individuals to remember events and to identify generic concerns.

COMMISSION FINDINGS

    d. I&E inspectors at various times have had difficulties having safety issues that they have raised seriously considered within the office. For example, in 1978 one I&E inspector raised the issue of operator termination of HPI during the September 1977 incident at Davis-Besse. For some 5 months, none of his efforts produced any action. He then took advantage of the "open-door policy" of NRC and went directly to two of the commissioners. These commissioners considered his complaint serious enough to merit further exploration. Unfortunately, this meeting with the commissioners did not take place until one week before the TMI-2 accident.

    e. Early this year, the General Accounting Office concluded that NRC had not made effective use of its authority to assess monetary penalties for significant violations. The report cited cases where I&E consolidated continuing violations into one violation, took too long to impose penalties, and sometimes reduced the penalties to avoid financial hardship for the licensee.

    f. In its investigative report on the TMI-2 accident (NUREG 0600), I&E came to the unequivocal conclusion that if the operators had followed their procedures for loss-of-coolant accidents, there would have been no accident. However, for more than 2 hours on March 28, the operators at TMI did not recognize that they had a loss-of-coolant accident and did not consider the LOCA procedure relevant. In any event, the TMI-2 procedures were inconsistent and misleading in this regard.

  10. There is an absence throughout the NRC of any overall system to measure and improve the quality of safety regulations. There are inadequate management and internal quality assurance systems, an inadequate research program, and the absence of any systematic effort to obtain and use the public health-related research of such federal agencies as HEW and EPA.

  11. The information and direction issued by NRC to licensees based on operating experience was, at times, fragmented and misleading. For example:

    a. An NRC publication describing the September 1977 Davis-Besse incident made no mention of the fact that operators interrupted HPI. The incident appeared under the heading of "valve malfunction" not "operator error."

    b. In the weeks following the accident, NRC apparently was confused as to what emergency procedures plant operators should follow. Thus, within a short span of time, NRC issued and then either modified or contradicted its post-TMI emergency instructions.

        (i) Immediately after the TMI accident, NRC directed operators not to override automatic engineered safety features under any circumstances and to operate high pressure injection without regard for reactor vessel pressure/temperature

COMMISSION FINDINGS

limits. NRC modified this directive within a short time.

(ii) On April 5, NRC required all licensees operating B&W-designed reactors to revise their procedures so that in the event of HPI initiation with reactor coolant pumps (RCP) operating, at least two RCPs would remain operating. On July 26, NRC took the opposite position and directed licensees to shut down its pumps when HPI initiated. I&E, in its August 1979 report on the TMI accident, stated that the failure of the TMI operators to shut down the RCPs sooner than they did was a potential item of noncompliance.

12. With its present organization, staff, and attitudes, the NRC is unable to fullfill its responsibility for providing an acceptable level of safety for nuclear power plants.

COMMISSION FINDINGS

H. THE PUBLIC'S RIGHT TO PUBLIC INFORMATION

1. The quality of information provided to the public in the event of a nuclear plant accident has a significant bearing on the capacity of people to respond to the accident, on their mental health, and on their willingness to accept guidance from responsible public officials.

2. Before the accident, Met Ed had consistently asserted the overall safety of the plant, although the company had made information concerning difficulties at TMI-2 public in weekly press releases. This information was not pursued, and often not understood, by the local news media in the area; and the local news media generally failed to publish or broadcast investigative stories on the safety of the plant.

3. Neither Met Ed nor the NRC had specific plans for providing accident information to the public and the news media.

4. During the accident, official sources of information were often confused or ignorant of the facts. News media coverage often reflected this confusion and ignorance.

5. Met Ed's handling of information during the first 3 days of the accident resulted in loss of its credibility as an information source with state and local officials, as well as with the news media. Part of the problem was that the utility was slow to confirm "pessimistic" news about the accident.

6. In accordance with an informal agreement worked out between Governor Thornburgh and the White House, the release of information was centralized beginning on the third day of the accident. Under the agreement, Harold Denton of the NRC would issue all statements from the site on plant status; the Governor's office would be the sole source of comment on protective action and evacuation; and the White House would coordinate comment on the federal emergency relief effort. This agreement limited the number of sources available to the news media and while it brought some

COMMISSION FINDINGS

order out of the chaos in public information, it raised two problems. First, information on off-site radiation releases was not centralized in any source so that it would be readily available to the news media and the public; and second, the plan provided no specific public information role for the utility.

7. During the first days of the accident, B&W made a conscious decision not to comment on the accident, even when company officials believed that misinformation was being made available to the public by others.

8. The reporters who covered the accident had widely divergent skills and backgrounds. Many had no scientific background. Because too few technical briefers were supplied by NRC and the utility, and because many reporters were unfamiliar with the technology and the limits of scientific knowledge, they had difficulty understanding fully the information that was given to them. In turn, the news media had difficulty presenting this information to the public in a form that would be understandable.

  a. This difficulty was particularly acute in the reporting of information on radiation releases.

  b. They also experienced difficulty interpreting language expressing the probability of such events as a meltdown or a hydrogen explosion; this was made even more difficult when the sources of information were themselves uncertain about the probabilities.

9. The impression exists that in news coverage of the accident, the news media presented a more alarming than reassuring view of events. Without attempting to assess how alarming the accident may _in fact_ have been, an analysis of the sources quoted in the news media reveals, overall, a larger proportion of reassuring than alarming statements in the coverage concerning the status of the accident. In choosing quotations from both official and unofficial sources, the news media did not present only "alarming" views, but rather views on both sides of issues related to the accident.

10. A qualitative survey of 42 newspapers from around the country showed that the vast majority covered the accident in much the same way as the major suppliers of news, such as the wire services, the broadcast networks, The New York Times, and The Washington Post. A few newspapers, however, did present a more frightening and misleading impression of the accident. This impression was created through headlines and graphics, and in the selection of material to print.

A. The Nuclear Regulatory Commission

B. The Utility and Its Suppliers

C. Training of Operating Personnel

D. Technical Assessment

E. Worker and Public Health and Safety

F. Emergency Planning and Response

G. The Public's Right to Information

COMMISSION RECOMMENDATIONS

A. THE NUCLEAR REGULATORY COMMISSION

The Commission found a number of inadequacies in the NRC and, therefore, proposes a restructuring of the agency. Because there is insufficient direction in the present statute, the President and Congress should consider incorporating many of the following measures in statutory form.

Agency Organization and Management

The Commission believes that as presently constituted, the NRC does not possess the organizational and management capabilities necessary for the effective pursuit of safety goals. The Commission recommends:

1. The Nuclear Regulatory Commission should be restructured as a new independent agency in the executive branch.

    a. The present five-member commission should be abolished.

    b. The new agency should be headed by a single administrator appointed by the President, subject to the advice and consent of the Senate, to serve a substantial term (not coterminous with that of the President) in order to provide an expectation of continuity, but at the pleasure of the President to allow removal when the President deems it necessary. The administrator should be a person from outside the present agency.

    c. The administrator should have substantial discretionary authority over the internal organization and management of the new agency, and over personnel transfers from the existing NRC. Unlike the present NRC arrangement, the administrator and major staff components should be located in the same building or group of buildings.

    d. A major role of the administrator should be assuring that offices within the agency communicate sufficiently so that research, operating experience, and inspection and enforcement affect the overall performance of the agency.

61

COMMISSION RECOMMENDATIONS

2. An oversight committee on nuclear reactor safety should be established. Its purpose would be to examine, on a continuing basis, the performance of the agency and of the nuclear industry in addressing and resolving important public safety issues associated with the construction and operation of nuclear power plants, and in exploring the overall risks of nuclear power.

a. The members of the committee, not to exceed 15 in number, should be appointed by the President and should include: persons conversant with public health, environmental protection, emergency planning, energy technology and policy, nuclear power generation, and nuclear safety; one or more state governors; and members of the general public.

b. The committee, assisted by its own staff, should report to the President and to Congress at least annually.

3. The Advisory Committee on Reactor Safeguards (ACRS) should be retained, in a strengthened role, to continue providing an independent technical check on safety matters. The members of the committee should continue to be part-time appointees; the Commission believes that the independence and high quality of the members might be compromised by making them full-time federal employees. The Commission recommends the following changes:

a. The staff of ACRS should be strengthened to provide increased capacity for independent analysis. Special consideration should be given to improving ACRS' capabilities in the field of public health.

b. The ACRS should not be required to review each license application. When ACRS chooses to review a license application, it should have the statutory right to intervene in hearings as a party. In particular, ACRS should be authorized to raise any safety issue in licensing proceedings, to give reasons and arguments for its views, and to require formal response by the agency to any submission it makes. Any member of ACRS should be authorized to appear and testify in hearings, but should be exempt from subpoena in any proceedings in which he has not previously appeared voluntarily or made an individual written submission.

c. ACRS should have similar rights in rulemaking proceedings. In particular, it should have the power to initiate a rulemaking proceeding before the agency to resolve any generic safety issue it identifies.

The Agency's Substantive Mandate

The new agency's primary statutory mission and first operating priority must be the assurance of safety in the generation of nuclear power, including safeguards of nuclear materials from theft, diversion, or loss. Accordingly, the Commission recommends the following:

COMMISSION RECOMMENDATIONS

    4. Included in the agency's general substantive charge should be the requirement to establish and explain safety-cost trade-offs; where additional safety improvements are not clearly outweighed by cost considerations, there should be a presumption in favor of the safety change. Transfers of statutory jurisdiction from the NRC should be preceded by a review to identify and remove any unnecessary responsibilities that are not germane to safety. There should also be emphasis on the relationship of the new agency's safety activities to related activities of other agencies. (See recommendations E.2 and F.1.b.)

    a. The agency should be directed to upgrade its operator and supervisor licensing functions. These should include the accreditation of training institutions from which candidates for a license must graduate. Such institutions should be required to employ qualified instructors, to perform emergency and simulator training, and to include instruction in basic principles of reactor science, reactor safety, and the hazards of radiation. The agency should also set criteria for operator qualifications and background investigations, and strictly test license candidates for the particular power plant they will operate. The agency should periodically review and reaccredit all training programs and relicense individuals on the basis of current information on experience in reactor operations. (See recommendations C.1 and C.2.)

    b. The agency should be directed to employ a broader definition of matters relating to safety that considers thoroughly the full range of safety matters, including, but not limited to, those now identified as "safety-related" items, which currently receive special attention.

    c. Other safety emphases should include:

    (i) a systems engineering examination of overall plant design and performance, including interaction among major systems and increased attention to the possibility of multiple failures;

    (ii) review and approval of control room design; the agency should consider the need for additional instrumentation and for changes in overall design to aid understanding of plant status, particularly for response to emergencies; (see recommendation D.1) and

    (iii) an increased safety research capacity with a broadly defined scope that includes issues relevant to public health. It is particularly necessary to coordinate research with the regulatory process in an effort to assure the maximum application of scientific knowledge in the nuclear power industry.

    5. Responsibility and accountability for safe power plant operations, including the management of a plant during an accident,

COMMISSION RECOMMENDATIONS

should be placed on the licensee in all circumstances. It is therefore necessary to assure that licensees are competent to discharge this responsibility. To assure this competency, and in light of our findings regarding Metropolitan Edison, we recommend that the agency establish and enforce higher organizational and management standards for licensees. Particular attention should be given to such matters as the following: integration of decision-making in any organization licensed to construct or operate a plant; kinds of expertise that must be within the organization; financial capability; quality assurance programs; operator and supervisor practices and their periodic reevaluation; plant surveillance and maintenance practices; and requirements for the analysis and reporting of unusual events.

6. In order to provide an added contribution to safety, the agency should be required, to the maximum extent feasible, to locate new power plants in areas remote from concentrations of population. Siting determinations should be based on technical assessments of various classes of accidents that can take place, including those involving releases of low doses of radiation. (See recommendation F.2.)

7. The agency should be directed to include, as part of its licensing requirements, plans for the mitigation of the consequences of accidents, including the cleanup and recovery of the contaminated plant. The agency should be directed to review existing licenses and to set deadlines for accomplishing any necessary modifications. (See recommendations D.2 and D.4.)

8. Because safety measures to afford better protection for the affected population can be drawn from the high standards for plant safety recommended in this report, the NRC or its successor should, on a case-by-case basis, before issuing a new construction permit or operating license:

    a. assess the need to introduce new safety improvements recommended in this report, and in NRC and industry studies;

    b. review, considering the recommendations set forth in this report, the competency of the prospective operating licensee to manage the plant and the adequacy of its training program for operating personnel; and

    c. condition licensing upon review and approval of the state and local emergency plans.

Agency Procedures

The Commission believes that the agency must improve on prior performance in resolving generic and specific safety issues. Generic safety issues are considered in rulemaking proceedings that formulate new standards for categories of plants. Specific safety issues are considered in adjudicative proceedings that determine whether a particular plant should receive a license. Both kinds of safety issues

COMMISSION RECOMMENDATIONS

are then dealt with in inspection and enforcement processes. The Commission believes that all of these agency functions need improvement, and accordingly recommends the following measures:

9. The agency's authorization to make general rules affecting safety should:

    a. require the development of a public agenda according to which rules will be formulated;

    b. require the agency to set deadlines for resolving generic safety issues;

    c. require a periodic and systematic reevaluation of the agency's existing rules; and

    d. define rulemaking procedures designed to create a process that provides a meaningful opportunity for participation by interested persons, that ensures careful consideration and explanation of rules adopted by the agency, and that includes appropriate provision for the application of new rules to existing plants. In particular, the agency should: accompany newly proposed rules with an analysis of the issues they raise and provide an indication of the technical materials that are relevant; provide a sufficient opportunity for interested persons to evaluate and rebut materials relied on by the agency or submitted by others; explain its final rules fully, including responses to principal comments by the public, the ACRS, and other agencies on proposed rules; impose when necessary special interim safeguards for operating plants affected by generic safety rulemaking; and conduct systematic reviews of operating plants to assess the need for retroactive application of new safety requirements.

10. Licensing procedures should foster early and meaningful resolution of safety issues before major financial commitments in construction can occur. In order to ensure that safety receives primary emphasis in licensing, and to eliminate repetitive consideration of some issues in that process, the Commission recommends the following:

    a. Duplicative consideration of issues in several stages of one plant's licensing should, wherever possible, be reduced by allocating particular issues (such as the need for power) to a single stage of the proceedings.

    b. Issues that recur in many licensings should be resolved by rulemaking.

    c. The agency should be authorized to conduct a combined construction permit and operating license hearing whenever plans can be made sufficiently complete at the construction permit stage.

    d. There should be provision for the initial adjudication of license applications and for appeal to a board whose decisions would not

COMMISSION RECOMMENDATIONS

be subject to further appeal to the administrator. Both initial adjudicators and appeal boards should have a clear mandate to pursue any safety issue, whether or not it is raised by a party.

  e. An Office of Hearing Counsel should be established in the agency. This office would not engage in the informal negotiations between other staff and applicants that typically precede formal hearings on construction permits. Instead, it would participate in the formal hearings as an objective party, seeking to assure that vital safety issues are addressed and resolved. The office should report directly to the administrator and should be empowered to appeal any adverse licensing board determination to the appeal board.

  f. Any specific safety issue left open in licensing proceedings should be resolved by a deadline.

 11. The agency's inspection and enforcement functions must receive increased emphasis and improved management, including the following elements:

  a. There should be an improved program for the systematic safety evaluation of currently operating plants, in order to assess compliance with current requirements, to assess the need to make new requirements retroactive to older plants, and to identify new safety issues.

  b. There should be a program for the systematic assessment of experience in operating reactors, with special emphasis on discovering patterns in abnormal occurrences. An overall quality assurance measurement and reporting system based on this systematic assessment shall be developed to provide: 1) a measure of the overall improvement or decline in safety, and 2) a base for specific programs aimed at curing deficiencies and improving safety. Licensees must receive clear instructions on reporting requirements and clear communications summarizing the lessons of experience at other reactors.

  c. The agency should be authorized and directed to assess substantial penalties for licensee failure to report new "safety-related" information or for violations of rules defining practices or conditions already known to be unsafe.

  d. The agency should be directed to require its enforcement personnel to perform improved inspection and auditing of licensee compliance with regulations and to conduct major and unannounced on-site inspections of particular plants.

  e. Each operating licensee should be subject periodically to intensive and open review of its performance according to the requirements of its license and applicable regulations.

COMMISSION RECOMMENDATIONS

    f. The agency should be directed to adopt criteria for revocation of licenses, sanctions short of revocation such as probationary status, and kinds of safety violations requiring immediate plant shutdown or other operational safeguards.

COMMISSION RECOMMENDATIONS

B. THE UTILITY AND ITS SUPPLIERS

1. To the extent that the industrial institutions we have examined are representative of the nuclear industry, the nuclear industry must dramatically change its attitudes toward safety and regulations. The Commission has recommended that the new regulatory agency prescribe strict standards. At the same time, the Commission recognizes that merely meeting the requirements of a government regulation does not guarantee safety. Therefore, the industry must also set and police its own standards of excellence to ensure the effective management and safe operation of nuclear power plants.

 a. The industry should establish a program that specifies appropriate safety standards including those for management, quality assurance, and operating procedures and practices, and that conducts independent evaluations. The recently created Institute of Nuclear Power Operations, or some similar organization, may be an appropriate vehicle for establishing and implementing this program.

 b. There must be a systematic gathering, review, and analysis of operating experience at all nuclear power plants coupled with an industry-wide international communications network to facilitate the speedy flow of this information to affected parties. If such experiences indicate the need for modifications in design or operation, such changes should be implemented according to realistic deadlines.

2. Although the Commission considers the responsibility for safety to be with the total organization of the plant, we recommend that each nuclear power plant company have a separate safety group that reports to high-level management. Its assignment would be to evaluate regularly procedures and general plant operations from a safety perspective; to assess quality assurance programs; and to develop continuing safety programs.

3. Integration of management responsibility at all levels must be achieved consistently throughout this industry. Although there may not be a single optimal management structure for nuclear power plant operation, there must be a single accountable organization with the requisite expertise to take responsibility for the integrated management

COMMISSION RECOMMENDATIONS

of the design, construction, operation, and emergency response functions, and the organizational entities that carry them out. Without such demonstrated competence, a power plant operating company should not qualify to receive an operating license.

    a. These goals may be obtained at the design stage by 1) contracting for a "turn-key" plant in which the vendor or architect-engineer contracts to supply a fully operational plant and supervises all planning, construction, and modification; or 2) assembling expertise capable of integrating the design process. In either case, it is critical that the knowledge and expertise gained during design and construction of the plant be effectively transferred to those responsible for operating the plant.

    b. Clearly defined roles and responsibilities for operating procedures and practices must be established to ensure accountability and smooth communication.

    c. Since, under our recommendations, accountability for operations during an emergency would rest on the licensee, the licensee must prepare clear procedures defining management roles and responsibilities in the event of a crisis.

4. It is important to attract highly qualified candidates for the positions of senior operator and operator supervisor. Pay scales should be high enough to attract such candidates.

5. Substantially more attention and care must be devoted to the writing, reviewing, and monitoring of plant procedures.

    a. The wording of procedures must be clear and concise.

    b. The content of procedures must reflect both engineering thinking and operating practicalities.

    c. The format of procedures, particularly those that deal with abnormal conditions and emergencies, must be especially clear, including clear diagnostic instructions for identifying the particular abnormal conditions confronting the operators.

    d. Management of both utilities and suppliers must insist on the early diagnosis and resolution of safety questions that arise in plant operations. They must also establish deadlines, impose sanctions for the failure to observe such deadlines, and make certain that the results of the diagnoses and any proposed procedural changes based on them are disseminated to those who need to know them.

6. Utility rate-making agencies should recognize that implementation of new safety measures can be inhibited by delay or failure to include the costs of such measures in the utility rate base. The Commission, therefore, recommends that state rate-making agencies give explicit attention to the safety implications of rate-making when they consider costs based on "safety-related" changes.

COMMISSION RECOMMENDATIONS

C. <u>TRAINING OF OPERATING PERSONNEL</u>

    1. The Commission recommends the establishment of agency-accredited training institutions for operators and immediate supervisors of operators. These institutions should have highly qualified instructors, who will maintain high standards, stress understanding of the fundamentals of nuclear power plants and the possible health effects of nuclear power, and who will train operators to respond to emergencies. (See recommendation A.4.a.)

        a. These institutions could be national, regional, or specific to individual nuclear steam systems.

        b. Reactor operators should be required to graduate from an accredited training institution. Exemption should be made only in cases where there is clear, documentary evidence that the candidate already has the equivalent training.

        c. The training institutions should be subject to periodic review and reaccreditation by the restructured NRC.

        d. Candidates for the training institute must meet entrance requirements geared to the curriculum.

    2. Individual utilities should be responsible for training operators who are graduates of accredited institutions in the specifics of operating a particular plant. These operators should be examined and licensed by the restructured NRC, both at their initial licensing and at the relicensing stage. In order to be licensed, operators must pass every portion of the examination. Supervisors of operators, at a minimum, should have the same training as operators.

    3. Training should not end when operators are given their licenses.

        a. Comprehensive ongoing training must be given on a regular basis to maintain operators' level of knowledge.

        b. Such training must be continuously integrated with operating experience.

COMMISSION RECOMMENDATIONS

        c. Emphasis must be placed on diagnosing and controlling complex transients and on the fundamental understanding of reactor safety.

        d. Each utility should have ready access to a control room simulator. Operators and supervisors should be required to train regularly on the simulator. The holding of operator licenses should be contingent on performance on the simulator.

    4. Research and development should be carried out on improving simulation and simulation systems: a) to establish and sustain a higher level of realism in the training of operators, including dealing with transients; and b) to improve the diagnostics and general knowledge of nuclear power plant systems.

COMMISSION RECOMMENDATIONS

D. TECHNICAL ASSESSMENT

1. Equipment should be reviewed from the point of view of providing information to operators to help them prevent accidents and to cope with accidents when they occur. Included might be instruments that can provide proper warning and diagnostic information; for example, the measurement of the full range of temperatures within the reactor vessel under normal and abnormal conditions, and indication of the actual position of valves. Computer technology should be used for the clear display for operators and shift supervisors of key measurements relevant to accident conditions, together with diagnostic warnings of conditions.

In the interim, consideration should be given to requiring, at TMI and similar plants, the grouping of these key measurements, including distinct warning signals on a single panel available to a specified operator and the providing of a duplicate panel of these key measurements and warnings in the shift supervisor's office.

2. Equipment design and maintenance inadequacies noted at TMI should be reviewed from the point of view of mitigating the consequences of accidents. Inadequacies noted in the following should be corrected: iodine filters, the hydrogen recombiner, the vent gas system, containment isolation, reading of water levels in the containment isolation, reading of water levels in the containment area, radiation monitoring in the containment building, and the capability to take and quickly analyze samples of containment atmosphere and water in various places. (See recommendation A.7.)

3. Monitoring instruments and recording equipment should be provided to record continuously all critical plant measurements and conditions.

4. The Commission recommends that continuing in-depth studies should be initiated on the probabilities and consequences (on-site and off-site) of nuclear power plant accidents, including the consequences of meltdown.

    a. These studies should include a variety of small-break loss-of-coolant accidents and multiple-failure accidents, with particular attention to human failures.

COMMISSION RECOMMENDATIONS

  b. Results of these studies should be used to help plan for recovery and cleanup following a major accident.

  c. From these studies may emerge desirable modifications in the design of plants that will help prevent accidents and mitigate their consequences. For example:

    (i) Consideration should be given to equipment that would facilitate the controlled safe venting of hydrogen gas from the reactor cooling system.

    (ii) Consideration should be given to overall gas-tight enclosure of the let-down/make-up system with the option of returning gases to the containment building.

  d. Such studies should be conducted by the industry and other qualified organizations and may be sponsored by the restructured NRC and other federal agencies.

 5. A study should be made of the chemical behavior and the extensive retention of radioactive iodine in water, which resulted in the very low release of radioiodine to the atmosphere in the TMI-2 accident. This information should be taken into account in the studies of the consequences of other small-break accidents.

 6. Since there are still health hazards associated with the cleanup and disposal process, which is being carried out for the first time in a commercial nuclear power plant, the Commission recommends close monitoring of the cleanup process at TMI and of the transportation and disposal of the large amount of radioactive material. As much data as possible should be preserved and recorded about the conditions within the containment building so that these may be used for future safety analyses.

 7. The Commission recommends that as a part of the formal safety assurance program, every accident or every new abnormal event be carefully screened, and where appropriate be rigorously investigated, to assess its implications for the existing system design, computer models of the system, equipment design and quality, operations, operator training, operator training simulators, plant procedures, safety systems, emergency measures, management, and regulatory requirements.

COMMISSION RECOMMENDATIONS

E. WORKER AND PUBLIC HEALTH AND SAFETY

1. The Commission recommends the establishment of expanded and better coordinated health-related radiation effects research. This research should include, but not be limited to:

    a. biological effects of low levels of ionizing radiation;

    b. acceptable levels of exposure to ionizing radiation for the general population and for workers;

    c. development of methods of monitoring and surveillance, including epidemiologic surveillance to monitor and determine the consequences of exposure to radiation of various population groups, including workers;

    d. development of approaches to mitigate adverse health effects of exposure to ionizing radiation; and

    e. genetic or environmental factors that predispose individuals to increased susceptibility to adverse effects.

This effort should be coordinated under the National Institutes of Health -- with an interagency committee of relevant federal agencies to establish the agenda for research efforts -- including the commitment of a portion of the research budget to meet the specific needs of the restructured NRC.

2. To ensure the best available review of radiation-related health issues, including reactor siting issues, policy statements or regulations in that area of the restructured NRC should be subject to mandatory review and comment by the Secretary of the Department of Health and Human Services. A time limit for the review should be established to assure such review is performed in an expeditious manner.

3. The Commission recommends, as a state and local responsibility, an increased program for educating health professionals and emergency response personnel in the vicinity of nuclear power plants.

COMMISSION RECOMMENDATIONS

4. Utilities must make sufficient advance preparation for the mitigation of emergencies:

    a. Radiation monitors should be available for monitoring of routine operations as well as accident levels.

    b. The emergency control center for health-physics operations and the analytical laboratory to be used in emergencies should be located in a well-shielded area supplied with uncontaminated air.

    c. There must be a sufficient health-related supply of instruments, respirators, and other necessary equipment for both routine and emergency conditions.

    d. There should be an adequate maintenance program for all such health-related equipment.

5. An adequate supply of the radiation protective (thyroid blocking) agent, potassium iodide for human use, should be available regionally for distribution to the general population and workers affected by a radiological emergency.

COMMISSION RECOMMENDATIONS

F. EMERGENCY PLANNING AND RESPONSE

1. Emergency plans must detail clearly and consistently the actions public officials and utilities should take in the event of off-site radiation doses resulting from release of radioactivity. Therefore, the Commission recommends that:

    a. Before a utility is granted an operating license for a new nuclear power plant, the state within which that plant is to be sited must have an emergency response plan reviewed and approved by the Federal Emergency Management Agency (FEMA). The agency should assess the criteria and procedures now used for evaluating state and local government plans and for determining their ability to activate the plans. FEMA must assure adequate provision, where necessary, for multi-state planning.

    b. The responsibility at the federal level for radiological emergency planning, including planning for coping with radiological releases, should rest with FEMA. In this process, FEMA should consult with other agencies, including the restructured NRC and the appropriate health and environmental agencies. (See recommendation A.4.)

    c. The state must effectively coordinate its planning with the utility and with local officials in the area where the plant is to be located.

    d. States with plants already operating must upgrade their plans to the requirements to be set by FEMA. Strict deadlines must be established to accomplish this goal.

2. Plans for protecting the public in the event of off-site radiation releases should be based on technical assessment of various classes of accidents that can take place at a given plant.

    a. No single plan based on a fixed set of distances and a fixed set of responses can be adequate. Planning should involve the identification of several different kinds of accidents with different possible radiation consequences. For each such scenario, there should be clearly identified criteria for the appropriate responses at various

COMMISSION RECOMMENDATIONS

distances, including instructing individuals to stay indoors for a period of time, providing special medication, or ordering an evacuation.

    b. Similarly, response plans should be keyed to various possible scenarios and activated when the nature and potential hazard of a given accident has been identified.

    c. Plans should exist for protecting the public at radiation levels lower than those currently used in NRC-prescribed plans.

    d. All local communities should have funds and technical support adequate for preparing the kinds of plans described above.

3. Research should be expanded on medical means of protecting the public against various levels and types of radiation. This research should include exploration of appropriate medications that can protect against or counteract radiation.

4. If emergency planning and response to a radiation-related emergency is to be effective, the public must be better informed about nuclear power. The Commission recommends a program to educate the public on how nuclear power plants operate, on radiation and its health effects, and on protective actions against radiation. Those who would be affected by such emergency planning must have clear information on actions they would be required to take in an emergency.

5. Commission studies suggest that decision-makers may have over-estimated the human costs, in injury and loss of life, in many mass evacuation situations. The Commission recommends study into the human costs of radiation-related mass evacuation and the extent, if any, to which the risks in radiation-related evacuations differ from other types of evacuations. Such studies should take into account the effects of improving emergency planning, public awareness of such planning, and costs involved in mass evacuations.

6. Plans for providing federal technical support, such as radiological monitoring, should clearly specify the responsibilities of the various support agencies and the procedures by which those agencies provide assistance. Existing plans for the provision of federal assistance, particularly the Interagency Radiological Assistance Plan and the various memoranda of understanding among the agencies, should be reexamined and revised by the appropriate federal authorities in the light of the experience of the TMI accident, to provide for better coordination and more efficient federal support capability.

COMMISSION RECOMMENDATIONS

G. THE PUBLIC'S RIGHT TO INFORMATION

1. Federal and state agencies, as well as the utility, should make adequate preparation for a systematic public information program so that in time of a radiation-related emergency, they can provide timely and accurate information to the news media and the public in a form that is understandable. There should be sufficient division of briefing responsibilities as well as availability of informed sources to reduce confused and inaccurate information. The Commission therefore recommends:

    a. Since the utility must be responsible for the management of the accident, it should also be primarily responsible for providing information on the status of the plant to the news media and to the public; but the restructured NRC should also play a supporting role and be available to provide background information and technical briefings.

    b. Since the state government is responsible for decisions concerning protective actions, including evacuations, a designated state agency should be charged with issuing all information on this subject. This agency is also charged with the development of and dissemination of accurate and timely information on off-site radiation doses resulting from releases of radioactivity. This information should be derived from appropriate sources. (See recommendation F.1.) This agency should also set up the machinery to keep local officials fully informed of developments and to coordinate briefings to discuss any federal involvement in evacuation matters.

2. The provision of accurate and timely information places special responsibilities on the official sources of this information. The effort must meet the needs of the news media for information but without compromising the ability of operational personnel to manage the accident. The Commission therefore recommends that:

    a. Those who brief the news media must have direct access to informed sources of information.

    b. Technical liaison people should be designated to inform the briefers and to serve as a resource for the news media.

COMMISSION RECOMMENDATIONS

        c. The primary official news sources should have plans for the prompt establishment of press centers reasonably close to the site. These must be properly equipped, have appropriate visual aids and reference materials, and be staffed with individuals who are knowledgeable in dealing with the news media. These press centers must be operational promptly upon the declaration of a general emergency or its equivalent.

    3. The coverage of nuclear emergencies places special responsibilities on the news media to provide accurate and timely information. The Commission therefore recommends that:

        a. All major media outlets (wire services, broadcast networks, news magazines, and metropolitan daily newspapers) hire and train specialists who have more than a passing familiarity with reactors and the language of radiation. All other news media, regardless of their size, located near nuclear power plants should attempt to acquire similar knowledge or make plans to secure it during an emergency.

        b. Reporters discipline themselves to place complex information in a context that is understandable to the public and that allows members of the public to make decisions regarding their health and safety.

        c. Reporters educate themselves to understand the pitfalls in interpreting answers to "what if" questions. Those covering an accident should have the ability to understand uncertainties expressed by sources of information and probabilities assigned to various possible dangers.

    4. State emergency plans should include provision for creation of local broadcast media networks for emergencies that will supply timely and accurate information. Arrangements should be made to make available knowledgeable briefers to go on the air to clear up rumors and explain conditions at the plant. Communications between state officials, the utility, and the network should be prearranged to handle the possibility of an evacuation announcement.

    5. The Commission recommends that the public in the vicinity of a nuclear power plant be routinely informed of local radiation measurements that depart appreciably from normal background radiation, whether from normal or abnormal operation of the nuclear power plant, from a radioactivity cleanup operation such as that at TMI-2, or from other sources.

ACCOUNT OF THE ACCIDENT
PROLOGUE

On Wednesday, March 28, 1979, 36 seconds after the hour of 4:00 a.m., several water pumps stopped working in the Unit 2 nuclear power plant on Three Mile Island, 10 miles southeast of Harrisburg, Pennsylvania.[1] Thus began the accident at Three Mile Island. In the minutes, hours, and days that followed, a series of events -- compounded by equipment failures, inappropriate procedures, and human errors and ignorance -- escalated into the worst crisis yet experienced by the nation's nuclear power industry.

The accident focused national and international attention on the nuclear facility at Three Mile Island and raised it to a place of prominence in the minds of hundreds of millions. For the people living in such communities as Royalton, Goldsboro, Middletown, Hummelstown, Hershey, and Harrisburg, the rumors, conflicting official statements, a lack of knowledge about radiation releases, the continuing possibility of mass evacuation, and the fear that a hydrogen bubble trapped inside a nuclear reactor might explode were real and immediate. Later, Theodore Gross, provost of the Capitol Campus of Pennsylvania State University located in Middletown a few miles from TMI, would tell the Commission:

> Never before have people been asked to live with such ambiguity. The TMI accident -- an accident we cannot see or taste or smell . . . is an accident that is invisible. I think the fact that it is invisible creates a sense of uncertainty and fright on the part of people that may well go beyond the reality of the accident itself.[2]

The reality of the accident, the realization that such an accident could actually occur, renewed and deepened the national debate over nuclear safety and the national policy of using nuclear reactors to generate electricity.

-0-0-0-

Three Mile Island is home to two nuclear power plants, TMI-1 and TMI-2. Together they have a generating capacity of 1,700 megawatts,

Photo at left: Goldsboro, Pennsylvania, March 28, 1979.

ACCOUNT OF THE ACCIDENT

Simulated fuel pellets. The actual pellets are molded uranium oxide and are stacked one atop another inside the fuel rods. Each pellet is about one inch tall and less than a half-inch wide.

ACCOUNT OF THE ACCIDENT

enough electricity to supply the needs of 300,000 homes. The two plants are owned jointly by Pennsylvania Electric Company, Jersey Central Power & Light Company, and Metropolitan Edison Company, and operated by Met Ed. These three companies are subsidiaries of General Public Utilities Corporation, an electric utility holding company headquartered in Parsippany, New Jersey.3/

Each TMI plant is powered by its nuclear reactor. A reactor's function in a commercial power plant is essentially simple -- to heat water. The hot water, in turn, produces steam, which drives a turbine that turns a generator to produce electricity. Nuclear reactors are a product of high technology. In recent years, nuclear facilities of generating capacity much larger than those of earlier years -- including TMI-1 and TMI-2 -- have gone into service.4/

A nuclear reactor generates heat as a result of nuclear fission, the splitting apart of an atomic nucleus, most often that of the heavy atom uranium. Each atom has a central core called a nucleus. The nuclei of atoms typically contain two types of particles tightly bound together: protons, which carry a positive charge, and neutrons, which have no charge. When a free neutron strikes the nucleus of a uranium atom, the nucleus splits apart. This splitting -- or fission -- produces two smaller radioactive atoms, energy, and free neutrons. Most of the energy is immediately converted to heat. The neutrons can strike other uranium nuclei, producing a chain reaction and continuing the fission process. Not all free neutrons split atomic nuclei. Some, for example, are captured by atomic nuclei. This is important, because some elements, such as boron or cadmium, are strong absorbers of neutrons and are used to control the rate of fission, or to shut off a chain reaction almost instantaneously.5/

Uranium fuels all nuclear reactors used commercially to generate electricity in the United States. At TMI-2, the reactor core holds some 100 tons of uranium. The uranium, in the form of uranium oxide, is molded into cylindrical pellets, each about an inch tall and less than half-an-inch wide. The pellets are stacked one atop another inside fuel rods. These thin tubes, each about 12 feet long, are made of Zircaloy-4, a zirconium alloy. This alloy shell -- called the "cladding" -- transfers heat well and allows most neutrons to pass through.6/

TMI-2's reactor contained 36,816 fuel rods -- 208 in each of its 177 fuel assemblies. A fuel assembly contains not only fuel rods, but space for cooling water to flow between the rods and tubes that may contain control rods or instruments to measure such things as the temperature inside the core. TMI-2's reactor has 52 tubes with instruments and 69 with control rods.7/

Control rods contain materials that are called "poisons" by the nuclear industry because they are strong absorbers of neutrons and shut off chain reactions. The absorbing materials in TMI-2's control rods are 80 percent silver, 15 percent indium, and 5 percent cadmium. When the control rods are all inserted in the core, fission is effectively blocked, as atomic nuclei absorb neutrons so that they cannot split other nuclei. A chain reaction is initiated by withdrawing

ACCOUNT OF THE ACCIDENT

the control rods. By varying the number of and the length to which the control rods are withdrawn, operators can control how much power a plant produces. The control rods are held up by magnetic clamps. In an emergency, the magnetic field is broken and the control rods, responding to gravity, drop immediately into the core to halt fission. This is called a "scram."

The nuclear reactors used in commercial power plants possess several important safety features. They are designed so that it is impossible for them to explode like an atomic bomb. The primary danger from nuclear power stations is the potential for the release of radioactive materials produced in the reactor core as the result of fission. These materials are normally contained within the fuel rods.

A fuel rod assembly, containing 208 individual fuel rods, being inserted into the core of TMI-1.

ACCOUNT OF THE ACCIDENT

Damage to the fuel rods can release radioactive material into the reactor's cooling water and this radioactive material might be released to the environment if the other barriers -- the reactor coolant system and containment building barriers -- are also breached.8/

A nuclear plant has three basic safety barriers, each designed to prevent the release of radiation. The first line of protection is the fuel rods themselves, which trap and hold radioactive materials produced in the uranium fuel pellets. The second barrier consists of the reactor vessel and the closed reactor coolant system loop. The TMI-2 reactor vessel, which holds the reactor core and its control rods, is a 40-foot high steel tank with walls 8-½ inches thick. This tank, in turn, is surrounded by two, separated concrete-and-steel shields, with a total thickness of up to 9-½ feet, which

---

View of TMI-1's core taken from about 50 feet above the core. Each fuel rod assembly fits into one of the squares in the core's grid.

ACCOUNT OF THE ACCIDENT

absorb radiation and neutrons emitted from the reactor core. Finally, all this is set inside the containment building, a 193-foot high, reinforced-concrete structure with walls 4 feet thick.9/

To supply the steam that runs the turbine, both plants at TMI rely on a type of steam supply system called a pressurized water reactor. This simply means that the water heated by the reactor is kept under high pressure, normally 2,155 pounds per square inch in the TMI-2 plant.

In normal operations, it is important in a pressurized water reactor that the water that is heated in the core remain below "saturation" -- that is, the temperature and pressure combination at which water boils and turns to steam. In an accident, steam formation

---

Schematic of the TMI-2 facility.

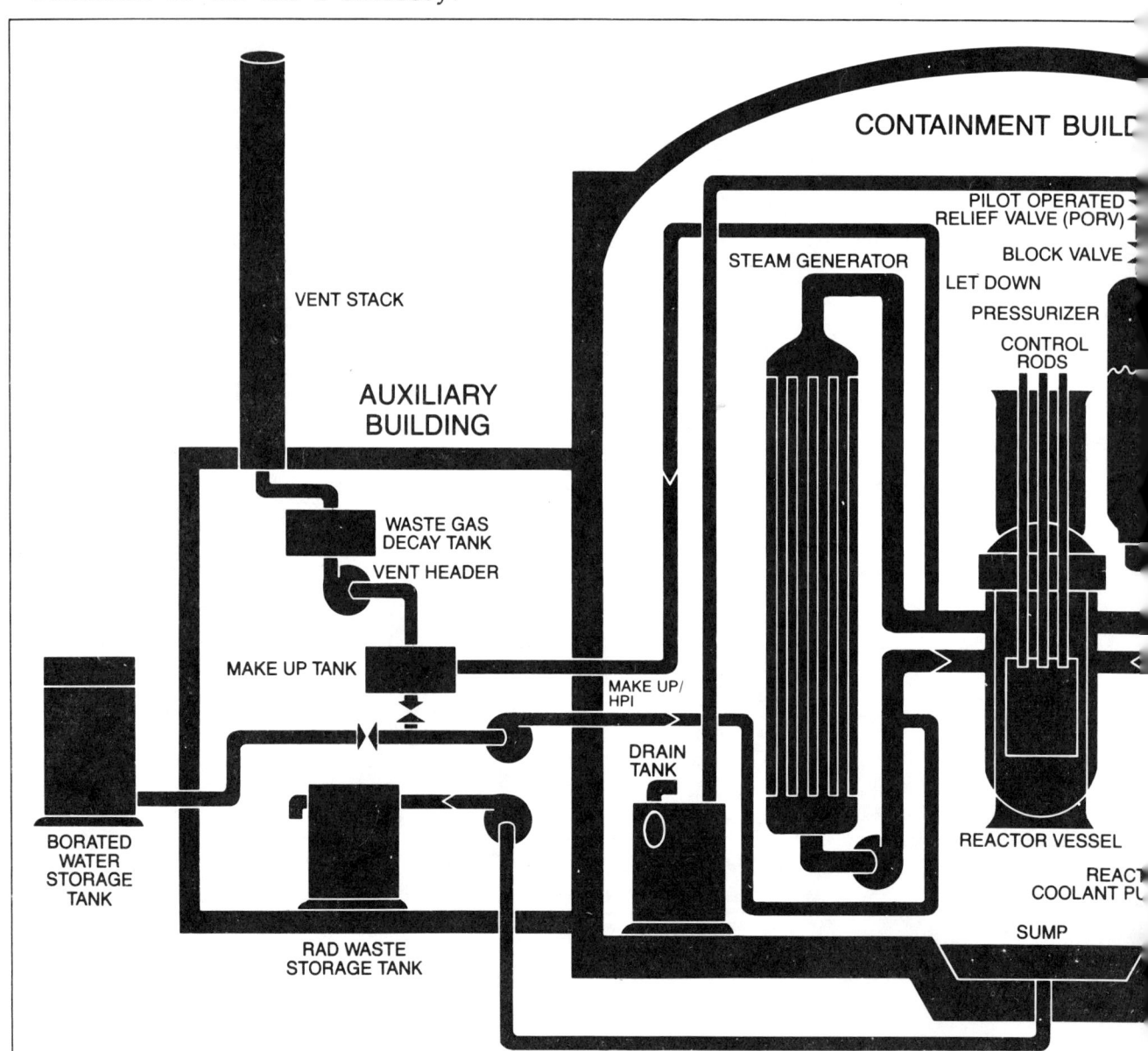

ACCOUNT OF THE ACCIDENT

itself is not a danger, because it too can help cool the fuel rods, although not as effectively as the coolant water. But problems can occur if so much of the core's coolant water boils away that the core becomes uncovered.

An uncovered core may lead to two problems. First, temperature may rise to a point, roughly 2,200°F, where a reaction of water and the cladding could begin to damage the fuel rods and also produce hydrogen. The other is that the temperature might rise above the melting point of the uranium fuel, which is about 5,200°F. Either poses a potential danger. Damage to the zirconium cladding releases some radioactive materials trapped inside the fuel rods into the core's cooling water. A melting of the fuel itself could release far more radioactive materials. If a significant portion of the

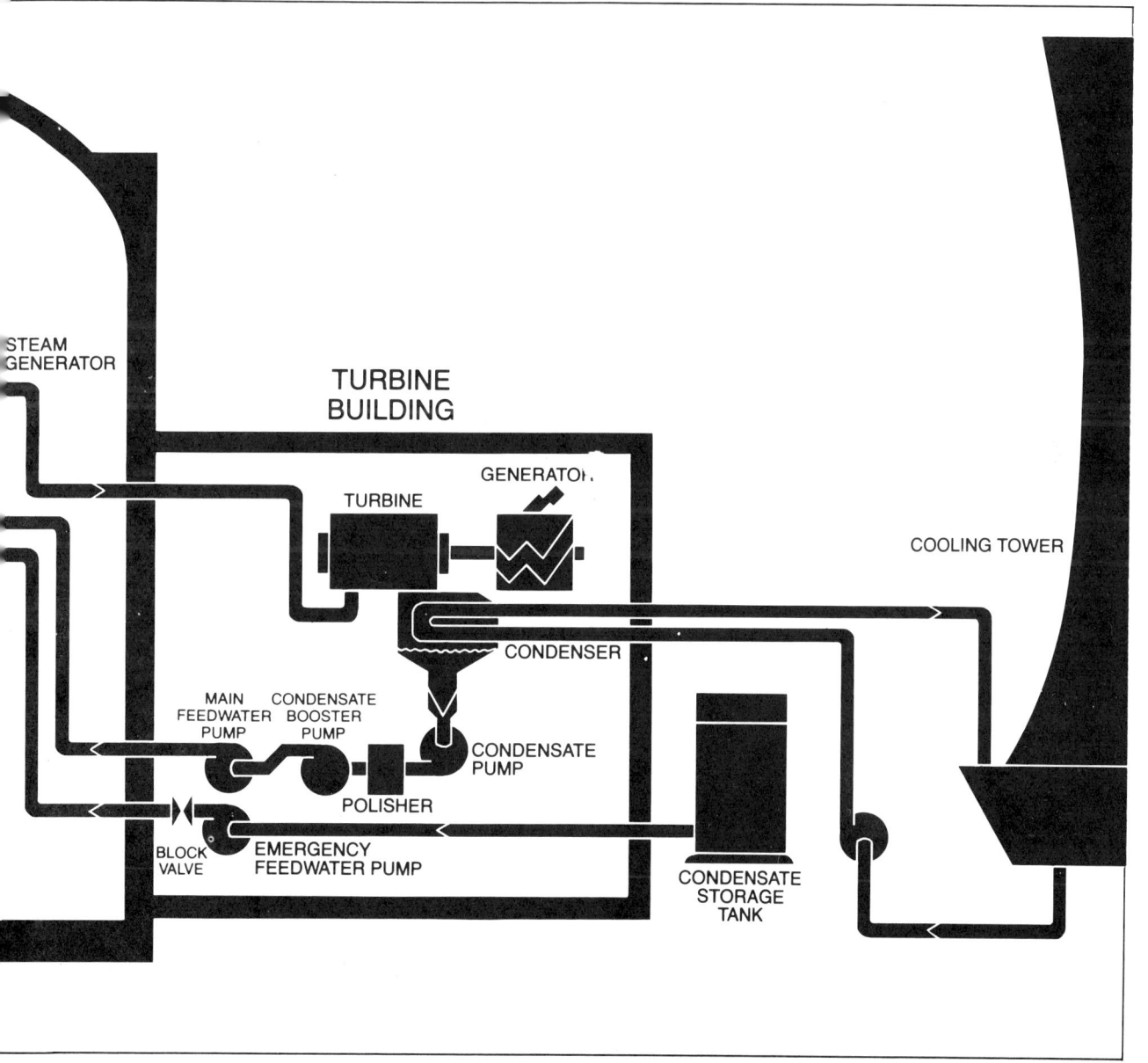

ACCOUNT OF THE ACCIDENT

fuel should melt, the molten fuel could melt through the reactor vessel itself and release large quantities of radioactive materials into the containment building. What might happen following such an event is very complicated and depends on a number of variables such as the specific characteristics of the materials on which a particular containment building is constructed.[10]

The essential elements of the TMI-2 system during normal operations include:

- The reactor, with its fuel rods and control rods.

- Water, which is heated by the fission process going on inside the fuel rods to ultimately produce steam to run the turbine. This water, by removing heat, also keeps the fuel rods from becoming overheated.

- Two steam generators, through which the heated water passes and gives up its heat to convert cooler water in another closed system to steam.

- A steam turbine that drives a generator to produce electricity.

- Pumps to circulate water through the various systems.

- A pressurizer, a large tank that maintains the reactor water at a pressure high enough to prevent boiling. At TMI-2, the pressurizer tank usually holds 800 cubic feet of water and 700 cubic feet of steam above it. The steam pressure is controlled by heating or cooling the water in the pressurizer. The steam pressure, in turn, is used to control the pressure of the water cooling the reactor.

Normally, water to the TMI-2 reactor flows through a closed system of pipes called the "reactor coolant system" or "primary loop." The water is pushed through the reactor by four reactor coolant pumps, each powered by a 9,000 horsepower electric motor. In the reactor, the water picks up heat as it flows around each fuel rod. Then it travels through 36-inch diameter, stainless steel pipes shaped like and called "candy canes," and into the steam generators.

In the steam generators, a transfer of heat takes place. The very hot water from the reactor coolant system travels down through the steam generators in a series of corrosion-resistant tubes. Meanwhile, water from another closed system -- the feedwater system or "secondary loop" -- is forced into the steam generator.

The feedwater in the steam generators flows around the tubes that contain the hot water from the reactor coolant system. Some of this heat is transferred to the cooler feedwater, which boils and becomes steam. Just as it would be in a coal- or oil-fired generating plant, the steam is carried from the two steam generators to turn the steam turbine, which runs the electricity-producing generator.

ACCOUNT OF THE ACCIDENT

The water from the reactor coolant system, which has now lost some of its heat, is pumped back to the reactor to pass around the fuel rods, pick up more heat, and begin its cycle again.

The water from the feedwater system, which has turned to steam to drive the turbine, passes through devices called condensers. Here, the steam is condensed back to water, and is forced back to the steam generators again.

The condenser water is cooled in the cooling towers. The water that cools the condensers is also in a closed system or loop. It cools the condensers, picks up heat, and is pumped to the cooling towers, where it cascades along a series of steps. As it does, it releases its heat to the outside air, creating the white vapor plumes that drift skyward from the towers. Then the water is pumped back to the condensers to begin its cooling process over again.

Neither the water that cools the condensers, nor the vapor plumes that rise from the cooling towers, nor any of the water that runs through the feedwater system is radioactive under normal conditions. The water that runs through the reactor coolant system is radioactive, of course, since it has been exposed to the radioactive materials in the core.

The turbine, the electric generator it powers, and most of the feedwater system piping are outside the containment building in other structures. The steam generators, however, which must be fed by water from both the reactor coolant and feedwater systems, are inside the containment building with the reactor and the pressurizer tank.

A nuclear power facility is designed with many ways to protect against system failure. Each of its major systems has an automatic backup system to replace it in the event of a failure. For example, in a loss-of-coolant accident (LOCA) -- that is, an accident in which there is a loss of the reactor's cooling water -- the Emergency Core Cooling System (ECCS) automatically uses existing plant equipment to ensure that cooling water covers the core.

In a LOCA, such as occurred at TMI-2, a vital part of the ECCS is the High Pressure Injection (HPI) pumps, which can pour about 1,000 gallons a minute into the core to replace cooling water being lost through a stuck-open valve, broken pipe, or other type of leak. But the ECCS can be effective only if plant operators allow it to keep running and functioning as designed. At Three Mile Island, they did not.

ACCOUNT OF THE ACCIDENT

WEDNESDAY, MARCH 28

In the parlance of the electric power industry, a "trip" means a piece of machinery stops operating. A series of feedwater system pumps supplying water to TMI-2's steam generators tripped on the morning of March 28, 1979. The nuclear plant was operating at 97 percent power at the time. The first pump trip occurred at 36 seconds after 4:00 a.m. When the pumps stopped, the flow of water to the steam generators stopped. With no feedwater being added, there soon would be no steam, so the plant's safety system automatically shut down the steam turbine and the electric generator it powered. The incident at Three Mile Island was 2 seconds old.

The production of steam is a critical function of a nuclear reactor. Not only does steam run the generator to produce electricity but also, as steam is produced, it removes some of the intense heat that the reactor water carries.

When the feedwater flow stopped, the temperature of the reactor coolant increased. The rapidly heating water expanded. The pressurizer level (the level of the water inside the pressurizer tank) rose and the steam in the top of the tank compressed. Pressure inside the pressurizer built to 2,255 pounds per square inch, 100 psi more than normal. Then a valve atop the pressurizer, called a pilot-operated relief valve, or PORV, opened -- as it was designed to do -- and steam and water began flowing out of the reactor coolant system through a drain pipe to a tank on the floor of the containment building.11/ Pressure continued to rise, however, and 8 seconds after the first pump tripped, TMI-2's reactor -- as it was designed to do -- scrammed: its control rods automatically dropped down into the reactor core to halt its nuclear fission.

Less than a second later, the heat generated by fission was essentially zero. But, as in any nuclear reactor, the decaying radioactive materials left from the fission process continued to heat the reactor's coolant water. This heat was a small fraction -- just 6 percent -- of that released during fission, but it was still substantial and had to be removed to keep the core from overheating. When the pumps that normally supply the steam generator with water

90

ACCOUNT OF THE ACCIDENT

shut down, three emergency feedwater pumps automatically started. Fourteen seconds into the accident, an operator in TMI-2's control room noted the emergency feed pumps were running. He did not notice two lights that told him a valve was closed on each of the two emergency feedwater lines and thus no water could reach the steam generators. One light was covered by a yellow maintenance tag. No one knows why the second light was missed. 12/

With the reactor scrammed and the PORV open, pressure in the reactor coolant system fell. Up to this point, the reactor system was responding normally to a turbine trip. The PORV should have closed 13 seconds into the accident, when pressure dropped to 2,205 psi. It did not. A light on the control room panel indicated that the electric power that opened the PORV had gone off, leading the operators to assume the valve had shut. 13/ But the PORV was stuck open, and would remain open for 2 hours and 22 minutes, draining needed coolant water -- a LOCA was in progress. In the first 100 minutes of the accident, some 32,000 gallons -- over one-third of the entire capacity of the reactor coolant system -- would escape through the PORV and out the reactor's let-down system. Had the valve closed as it was designed to do, or if the control room operators had realized that the valve was stuck open and closed a backup valve to stem the flow of coolant water, or if they had simply left on the plant's high pressure injection pumps, the accident at Three Mile Island would have remained little more than a minor inconvenience for Met Ed.

-0-0-0-

To a casual visitor, the control room at TMI-2 can be an intimidating place, with messages coming from the loudspeaker of the plant's paging system; panel upon panel of red, green, amber, and white lights; and alarms that sound or flash warnings many times each hour. Reactor operators are trained how to respond and to respond quickly in emergencies. Initial actions are ingrained, almost automatic and unthinking. 14/

The burden of dealing with the early, crucial stages of the accident at Three Mile Island fell to four men -- William Zewe, shift supervisor in charge of both TMI-1 and TMI-2; Fred Scheimann, shift foreman for TMI-2; and two control room operators, Edward Frederick and Craig Faust. Each had been trained for his job by Met Ed and Babcock & Wilcox, the company that supplied TMI-2's reactor and nuclear steam system; each was licensed by the Nuclear Regulatory Commission; each was a product of his training -- training that did not adequately prepare them to cope with the accident at TMI-2. 15/ Indeed, their training was partly responsible for escalating what should have been a minor event into a potentially devastating accident.

Frederick and Faust were in the control room 16/ when the first alarm sounded, followed by a cascade of alarms that numbered 100 within minutes. The operators reacted quickly as trained to counter the turbine trip and reactor scram. Later Faust would recall for the Commission his reaction to the incessant alarms: "I would have liked to have thrown away the alarm panel. It wasn't giving us any

ACCOUNT OF THE ACCIDENT

TMI-2 control room.

ACCOUNT OF THE ACCIDENT

useful information."17/ Zewe, working in a small, glass-enclosed office behind the operators, alerted the TMI-1 control room of the TMI-2 scram and called his shift foreman back to the control room.

Scheimann had been overseeing maintenance on the plant's Number 7 polisher -- one of the machines that remove dissolved minerals from the feedwater system. His crew was using a mixture of air and water to break up resin that had clogged a resin transfer line. Later investigation would reveal that a faulty valve in one of the polishers allowed some water to leak into the air-controlled system that opens and closes the polishers' valves and may have been a factor in their sudden closure just before the accident began. This malfunction probably triggered the initial pump trip that led to the accident. The same problem of water leaking into the polishers' valve control system had occurred at least twice before at TMI-2. Had Met Ed corrected the earlier polisher problem, the March 28 sequence of events may never have begun.18/

-0-0-0-

With the PORV stuck open and heat being removed by the steam generators, the pressure and temperature of the reactor coolant system dropped. The water level also fell in the pressurizer. Thirteen seconds into the accident, the operators turned on a pump to add water to the system. This was done because the water in the system was shrinking as it cooled. Thus more water was needed to fill the system. Forty-eight seconds into the incident, while pressure continued falling, the water level in the pressurizer began to rise again. The reason, at this point, was that the amount of water being pumped into the system was greater than that being lost through the PORV.

About a minute and 45 seconds into the incident, because their emergency water lines were blocked, the steam generators boiled dry. After the steam generators boiled dry, the reactor coolant heated up again, expanded, and this helped send the pressurizer level up further.

Two minutes into the incident, with the pressurizer level still rising, pressure in the reactor coolant system dropped sharply. Automatically, two large pumps began pouring about 1,000 gallons a minute into the system. The pumps, called high pressure injection (HPI) pumps, are part of the reactor's emergency core cooling system. The level of water in the pressurizer continued to rise, and the operators, conditioned to maintain a certain level in the pressurizer, took this to mean that the system had plenty of water in it.19/ However, the pressure of reactor coolant system water was falling, and its temperature became constant.

About 2-½ minutes after the HPI pumps began working, Frederick shut one down and reduced the flow of the second to less than 100 gallons per minute. The falling pressure, coupled with a constant reactor coolant temperature after HPI came on, should have clearly alerted the operators that TMI-2 had suffered a LOCA, and safety required they maintain high pressure injection. "The rapidly

ACCOUNT OF THE ACCIDENT

increasing pressurizer level at the onset of the accident led me to believe that the high pressure injection was excessive, and that we were soon going to have a solid system," Frederick later told the Commission.[20]

A solid system is one in which the entire reactor and its cooling system, including the pressurizer, are filled with water. The operators had been taught to keep the system from "going solid" -- a condition that would make controlling the pressure within the reactor coolant system more difficult and that might damage the system. The operators followed this line of reasoning, oblivious for over 4 hours to a far greater threat -- that the loss of water from the system could result in uncovering the core.[21]

The saturation point was reached 5-½ minutes into the accident. Steam bubbles began forming in the reactor coolant system, displacing the coolant water in the reactor itself. The displaced water moved into the pressurizer, sending its level still higher. This continued to suggest to the operators that there was plenty of water in the system. They did not realize that water was actually flashing into steam in the reactor, and with more water leaving the system than being added, the core was on its way to being uncovered.[22] And so the operators began draining off the reactor's cooling water through piping called the let-down system.

Eight minutes into the accident, someone -- just who is a matter of dispute -- discovered that no emergency feedwater was reaching the steam generators. Operator Faust scanned the lights on the control panel that indicate whether the emergency feedwater valves are open or closed. He first checked a set of emergency feedwater valves designed to open after the pumps reach full speed; they were open. Next he checked a second pair of emergency feedwater valves, called the "twelve-valves," which are always supposed to be open, except during a specific test of the emergency feedwater pumps. The two "twelve-valves" were closed. Faust opened them and water rushed into the steam generators.[23]

The two "twelve-valves" were known to have been closed 2 days earlier, on March 26, as part of a routine test of the emergency feedwater pumps. A Commission investigation has not identified a specific reason as to why the valves were closed at 8 minutes into the accident. The most likely explanations are: the valves were never reopened after the March 26 test; or the valves were reopened and the control room operators mistakenly closed the valves during the very first part of the accident; or the valves were closed mistakenly from control points outside the control room after the test. The loss of emergency feedwater for 8 minutes had no significant effect on the outcome of the accident.[24] But it did add to the confusion that distracted the operators as they sought to understand the cause of their primary problem.

Throughout the first 2 hours of the accident, the operators ignored or failed to recognize the significance of several things that should have warned them that they had an open PORV and a loss-of-coolant accident. One was the high temperatures at the

ACCOUNT OF THE ACCIDENT

TMI-2 control room operators testifying before the Commission.
Seated (l to r) are Ernest Blake, legal counsel to Met Ed, and operators
Fred Scheimann, William Zewe, Edward Frederick, and Craig Faust.

ACCOUNT OF THE ACCIDENT

drain pipe that led from the PORV to the reactor coolant drain tank. One emergency procedure states that a pipe temperature of 200°F indicates an open PORV. Another states that when the drain pipe temperature reaches 130°F, the block valve beneath it should be closed.25/ But the operators testified that the pipe temperature normally registered high because either the PORV or some other valve was leaking slightly. "I have seen, in reviewing logs since the accident, approximately 198 degrees," Zewe told the Commission. "But I can remember instances before . . . just over 200 degrees."26/ So Zewe and his crew dismissed the significance of the temperature readings, which Zewe recalled as being in the 230°F range. Recorded data show the range reached 285°F. Zewe told the Commission that he regarded the high temperatures on the drain pipe as residual heat: ". . .[K]nowing that the relief valve had lifted, the downstream temperature I would expect to be high and that it would take some time for the pipe to cool down below the 200-degree set point."27/

-0-0-0-

At 4:11 a.m., an alarm signaled high water in the containment building's sump, a clear indication of a leak or break in the system. The water, mixed with steam, had come from the open PORV, first falling to the drain tank on the containment building floor and finally filling the tank and flowing into the sump. At 4:15 a.m., a rupture disc on the drain tank burst as pressure in the tank rose. This sent more slightly radioactive water onto the floor and into the sump. From the sump it was pumped to a tank in the nearby auxiliary building.

Five minutes later, at 4:20 a.m., instruments measuring the neutrons inside the core showed a count higher than normal, another indication -- unrecognized by the operators -- that steam bubbles were present in the core and forcing cooling water away from the fuel rods. During this time, the temperature and pressure inside the containment building rose rapidly from the heat and steam escaping via the PORV and drain tank. The operators turned on the cooling equipment and fans inside the containment building. The fact that they failed to realize that these conditions resulted from a LOCA indicates a severe deficiency in their training to identify the symptoms of such an accident.28/

About this time, Edward Frederick took a call from the auxiliary building. He was told an instrument there indicated more than 6 feet of water in the containment building sump. Frederick queried the control room computer and got the same answer. Frederick recommended shutting off the two sump pumps in the containment building. He did not know where the water was coming from and did not want to pump water of unknown origin, which might be radioactive, outside the containment building.29/ Both sump pumps were stopped about 4:39 a.m. Before they were, however, as much as 8,000 gallons of slightly radioactive water may have been pumped into the auxiliary building.30/ Only 39 minutes had passed since the start of the accident.

ACCOUNT OF THE ACCIDENT

Control panel of TMI-2, showing maintenance tags that operators testified covered one of the closed emergency feedwater valve indicator lights during the first 8 minutes of the accident.

ACCOUNT OF THE ACCIDENT

Gary Miller (l), station manager of TMI, and George Kunder (r), TMI-2's superintendent for technical support, testifying before the Commission.

ACCOUNT OF THE ACCIDENT

-0-0-0-

George Kunder, superintendent of technical support at TMI-2, arrived at the Island about 4:45 a.m., summoned by telephone. Kunder was duty officer that day, and he had been told TMI-2 had had a turbine trip and reactor scram. What he found upon his arrival was not what he expected. "I felt we were experiencing a very unusual situation, because I had never seen pressurizer level go high and peg in the high range, and at the same time, pressure being low," he told the Commission. "They have always performed consistently."31/ Kunder's view was shared by the control room crew. They later described the accident as a combination of events they had never experienced, either in operating the plant or in their training simulations.32/

Shortly after 5:00 a.m., TMI-2's four reactor coolant pumps began vibrating severely. This resulted from pumping steam as well as water, and it was another indication that went unrecognized that the reactor's water was boiling into steam. The operators feared the violent shaking might damage the pumps -- which force water to circulate through the core -- or the coolant piping.33/

Zewe and his operators followed their training. At 5:14 a.m., two of the pumps were shut down. Twenty-seven minutes later, operators turned off the two remaining pumps, stopping the forced flow of cooling water through the core.

There was already evidence by approximately 6:00 a.m. that at least a few of the reactor's fuel rod claddings had ruptured from high gas pressures inside them, allowing some of the radioactive gases within the rods to escape into the coolant water. The early warning came from radiation alarms inside the containment building. With coolant continuing to stream out the open PORV and little water being added, the top of the core became uncovered and heated to the point where the zirconium alloy of the fuel rod cladding reacted with steam to produce hydrogen. Some of this hydrogen escaped into the containment building through the open PORV and drain tank; some of it remained within the reactor. This hydrogen, and possibly hydrogen produced later in the day, caused the explosion in the containment building on Wednesday afternoon and formed the gas bubble that produced such great concern a few days later.34/

Other TMI officials now were arriving in the TMI-2 control room. They included Richard Dubiel, a health physicist who served as supervisor of radiation protection and chemistry; Joseph Logan, superintendent of TMI-2; and Michael Ross, supervisor of operations for TMI-1.

Shortly after 6:00 a.m., George Kunder participated in a telephone conference call with John Herbein, Met Ed's vice president for generation; Gary Miller, TMI station manager and Met Ed's senior executive stationed at the nuclear facility; and Leland Rogers, the Babcock & Wilcox site representative at TMI. The four men discussed the situation at the plant. In his deposition, Rogers recalled a significant question he posed during that call: He asked if the block valve between the pressurizer and the PORV, a backup valve

ACCOUNT OF THE ACCIDENT

that could be closed if the PORV stuck open, had been shut.

QUESTION: What was the response?

ROGERS: George's immediate response was, "I don't know," and he had someone standing next to the shift supervisor over back of the control room and sent the guy to find out if the valve block was shut.

QUESTION: You heard him give these instructions?

ROGERS: Yes, and very shortly I heard the answer come back from the other person to George, and he said, "Yes, the block valve was shut. . . ."35/

The operators shut the block valve at 6:22 a.m., 2 hours and 22 minutes after the PORV had opened.

It remains, however, an open question whether Rogers or someone else was responsible for the valve being closed. Edward Frederick testified that the valve was closed at the suggestion of a shift supervisor coming onto the next shift; but Frederick has also testified that the valve was closed because he and his fellow operators could think of nothing else to do to bring the reactor back under control.36/

In any event, the loss of coolant was stopped, and pressure began to rise, but the damage continued. Evidence now indicates the water in the reactor was below the top of the core at 6:15 a.m. Yet for some unexplained reason, high pressure injection to replace the water lost through the PORV and let-down system was not initiated for almost another hour. Before that occurred, Kunder, Dubiel, and their colleagues would realize they faced a serious emergency at TMI-2.

In the 2 hours after the turbine trip, periodic alarms warned of low-level radiation within the unoccupied containment building. After 6:00 a.m., the radiation readings markedly increased. About 6:30 a.m., a radiation technician began surveying the TMI-2 auxiliary building, using a portable detector -- a task that took about 20 minutes. He reported rapidly increasing levels of radiation, up to one rem per hour. During this period, monitors in the containment and auxiliary buildings showed rising radiation levels. By 6:48 a.m., high radiation levels existed in several areas of the plant, and evidence indicates as much as two-thirds of the 12-foot high core stood uncovered at this time. Analyses and calculations made after the accident indicate temperatures as high as 3,500 to 4,000°F or more in parts of the core existed during its maximum uncovery.37/ At 6:54 a.m., the operators turned on one of the reactor coolant pumps, but shut it down 19 minutes later because of high vibrations. More radiation alarms went off. Shortly before 7:00 a.m., Kunder and Zewe declared a site emergency, required by TMI's emergency plan whenever some event threatens "an uncontrolled release of radioactivity to the immediate environment."38/

-0-0-0-

ACCOUNT OF THE ACCIDENT

Gary Miller, TMI station manager, arrived at the TMI-2 control room a few minutes after 7:00 a.m. Radiation levels were increasing throughout the plant. Miller had first learned of the turbine trip and reactor scram within minutes after they occurred. He had had several telephone conversations with people at the site, including the 6:00 a.m. conference call. When he reached Three Mile Island, Miller found that a site emergency existed. He immediately assumed command as emergency director and formed a team of senior employees to aid him in controlling the accident and in implementing TMI-2's emergency plan.39/

Miller told Michael Ross to supervise operator activities in the TMI-2 control room. Richard Dubiel directed radiation activities, including monitorings on- and off-site. Joseph Logan was charged with ensuring that all required procedures and plans were reviewed and followed. George Kunder took over technical support and communications. Daniel Shovlin, TMI's maintenance superintendent, directed emergency maintenance. B&W's Leland Rogers was asked to provide technical assistance and serve as liaison with his home office. Miller gave James Seelinger, superintendent of TMI-1, charge of the emergency control station set up in the TMI-1 control room.40/ Under TMI's emergency plan, the control room of the unit not involved in an accident becomes the emergency control station. On March 28, TMI-1 was in the process of starting again after being shut down for refueling of its reactor.

TMI personnel were already following the emergency plan, telephoning state authorities about the site emergency.41/ The Pennsylvania Emergency Management Agency (PEMA) was asked to notify the Bureau of Radiation Protection (BRP), part of Pennsylvania's Department of Environmental Resources. The bureau in turn telephoned Kevin Molloy, director of the Dauphin County Office of Emergency Preparedness. Dauphin County includes Harrisburg and Three Mile Island. Other nearby counties and the State Police were alerted.

Met Ed alerted the U.S. Department of Energy's Radiological Assistance Plan office at Brookhaven National Laboratory. But notifying the Nuclear Regulatory Commission's Region I office in King of Prussia, Pennsylvania, took longer. The initial phone call reached an answering service, which tried to telephone the NRC duty officer and the region's deputy director at their homes. Both were en route to work.

By the time the NRC learned of the accident -- when its Region I office opened at 7:45 a.m. -- Miller had escalated the site emergency at Three Mile Island to a general emergency. Shortly after 7:15 a.m., emergency workers had to evacuate the TMI-2 auxiliary building. William Dornsife, a nuclear engineer with the Pennsylvania Bureau of Radiation Protection, was on the telephone to the TMI-2 control room at the time. He heard the evacuation ordered over the plant's paging system. "And I said to myself, 'This is the biggie,'" Dornsife recalled in his deposition.42/

At 7:20 a.m., an alarm indicated that the radiation dome monitor high in the containment building was reading 8 rems per hour. The

ACCOUNT OF THE ACCIDENT

monitor is shielded by lead. This shielding is designed to cut the radioactivity reaching the monitor by 100 times. Thus, those in the control room interpreted the monitor's alarm as meaning that the radiation present in the containment building at the time was about 800 rems per hour. Almost simultaneously, the operators finally turned on the high pressure injection pumps, once again dumping water into the reactor, but this intense flow was kept on for only 18 minutes. Other radiation alarms sounded in the control room. Gary Miller declared a general emergency at 7:24 a.m. By definition, at Three Mile Island, a general emergency is an "incident which has the potential for serious radiological consequences to the health and safety of the general public."43/

As part of TMI's emergency plan, state authorities were again notified and teams were sent to monitor radiation on the Island and ashore. The first team, designated Alpha and consisting of two radiation technicians, was sent to the west side of the Island, the downwind direction at the time. Another two-man team, designated Charlie, left for Goldsboro, a community of some 600 persons on the west bank of the Susquehanna River across from Three Mile Island. Meanwhile, a team sent into the auxiliary building reported increasing radiation levels and the building's basement partly flooded with water. At 7:48 a.m., radiation team Alpha reported radiation levels along the Island's west shoreline were less than one millirem per hour. Minutes later, another radiation team reported similar readings at the Island's north gate and along Route 441, which runs parallel to the Susquehanna's eastern shore.

-0-0-0-

Nearly 4 hours after the accident began, the containment building automatically isolated. Isolation is intended to help prevent radioactive material released by an accident from escaping into the environment. The building is not totally closed off. Pipes carrying coolant run between the containment and auxiliary buildings. These pipes close off when the containment building isolates, but the operators can open them. This occurred at TMI-2 and radioactive water flowed through these pipes even during isolation. Some of this piping leaked radioactive material into the auxiliary building, some of which escaped from there into the atmosphere outside.44/

In September 1975, the NRC instituted its Standard Review Plan, which included new criteria for isolation. The plan listed three conditions -- increased pressure, rising radiation levels, and emergency core cooling system activation -- and required that containment buildings isolate on any two of the three. However, the plan was not applied to nuclear plants that had already received their construction permits. TMI-2 had, so it was "grandfathered" and not required to meet the Standard Review Plan, although the plant had yet to receive its operating license.45/

In the TMI-2 design, isolation occurred only when increasing pressure in the containment building reached a certain point, nominally 4 pounds per square inch. Radiation releases alone, no matter how intense, would not initiate isolation, nor would ECCS activation.46/

ACCOUNT OF THE ACCIDENT

Although large amounts of steam entered the containment building early in the TMI-2 accident through the open PORV, the operators had kept pressure there low by using the building's cooling and ventilation system. But the failure to isolate early made little difference in the TMI-2 accident. Some of the radioactivity ultimately released into the atmosphere occurred after isolation from leaks in the let-down system that continued to carry radioactive water out of the containment building into the auxiliary building.47/

At 8:26 a.m., the operators once again turned on the ECCS's high pressure injection pumps and maintained a relatively high rate of flow. The core was still uncovered at this time and evidence indicates it took until about 10:30 a.m. for the HPI pumps to fully cover the core again.

By 7:50 a.m., NRC Region I officials had established direct telephone contact with the TMI-2 control room. Ten minutes later, Region I activated its Incident Response Center at King of Prussia, opened a direct telephone line to the Emergency Control Station in the TMI-1 control room, and notified NRC staff headquarters in Bethesda, Maryland. Region I officials gathered what information they could and relayed it to NRC headquarters, which had activated its own Incident Response Center. Region I dispatched two teams of inspectors to Three Mile Island; the first left at about 8:45 a.m., the second a few minutes later.

Around 8:00 a.m., it was clear to Gary Miller that the TMI-2 reactor had suffered some fuel damage. The radiation levels told him that. Yet Miller would testify to the Commission: ". . . I don't believe in my mind I really believed the core had been totally uncovered, or uncovered to a substantial degree at that time."48/

Off the Island, radiation readings continued to be encouragingly low. Survey team Charlie reported no detectable radiation in Goldsboro. Miller and several aides concluded about 8:30 a.m. that the emergency plan was being properly implemented.

-0-0-0-

WKBO, a Harrisburg "Top 40" music station, broke the story of TMI-2 on its 8:25 a.m. newscast. The station's traffic reporter, known as Captain Dave, uses an automobile equipped with a CB radio to gather his information. About 8:00 a.m. he heard police and fire fighters were mobilizing in Middletown and relayed this to his station. Mike Pintek, WKBO's news director, called Three Mile Island and asked for a public relations official. He was connected instead with the control room to a man who told him: "I can't talk now, we've got a problem." The man denied that "there are any fire engines," and told Pintek to telephone Met Ed's headquarters in Reading, Pennsylvania.

Pintek did, and finally reached Blaine Fabian, the company's manager of communications services. In an interview with the Commission staff, Pintek told what happened next:

ACCOUNT OF THE ACCIDENT

> Fabian came on and said there was a general emergency. What the hell is that? He said that general emergency was a "red-tape" type of thing required by the NRC when certain conditions exist. What conditions? "There was a problem with a feedwater pump. The plant is shut down. We're working on it. There's no danger off-site. No danger to the general public." And that is the story we went with at 8:25. I tried to tone it down so people wouldn't be alarmed. 49/

At 9:06 a.m., the Associated Press filed its first story -- a brief dispatch teletyped to newspaper, television, and radio news rooms across the nation. The article quoted Pennsylvania State Police as saying a general emergency had been declared, "there is no radiation leak," and that Met Ed officials had requested a State Police helicopter "that will carry a monitoring team." The story contained only six sentences in four paragraphs, but it alerted editors to what would become one of the most heavily reported news stories of 1979. 50/

Many public officials learned of the accident from the news media, rather than from the state, or their own emergency preparedness people. Harrisburg Mayor Paul Doutrich was one, and that still rankled him when he testified before the Commission 7 weeks later. Doutrich heard about the problem in a 9:15 a.m. telephone call from a radio station in Boston. "They asked me what we were doing about the nuclear emergency," Doutrich recalled. "My response was, 'What nuclear emergency?' They said, 'Well, at Three Mile Island.' I said, 'I know nothing about it. We have a nuclear plant there, but I know nothing about a problem.' So they told me; a Boston radio station." 51/

-0-0-0-

At 9:15 a.m., the NRC notified the White House of the events at Three Mile Island. Seven minutes later, an air sample taken in Goldsboro detected low levels of radioactive iodine-131. This specific reading was erroneous; a later, more sensitive analysis of the sample found no iodine-131. At 9:30 a.m., John Herbein, Met Ed's vice president for generation, was ordered to Three Mile Island from Philadelphia by Met Ed President Walter Creitz. And at 10:05 a.m., the first contingent of NRC Region I officials arrived at Three Mile Island.

In the days to follow, the NRC would dominate the public's perception of the events at Three Mile Island. But the initial NRC team consisted of only five Region I inspectors, headed by Charles Gallina. The five were briefed in the TMI-1 control room on the status of TMI-2. Then Gallina sent two inspectors into the TMI-2 control room and two more out to take radiation measurements; he himself remained in the TMI-1 control room to coordinate their reports and relay information to both Region I and NRC headquarters. 52/

While the NRC team received its briefing, monitors indicated that radiation levels in the TMI-2 control room had risen above the levels considered acceptable in NRC regulations. Workers put on

ACCOUNT OF THE ACCIDENT

John Herbein (l), Met Ed vice president for generation, and
Walter Creitz (r), Met Ed president, at a March 29 press conference.

ACCOUNT OF THE ACCIDENT

protective face masks with filters to screen out any airborne radioactive particles. This made communications among those managing the accident difficult. At 11:00 a.m., all nonessential personnel were ordered off the Island. At the same hour, both Pennsylvania's Bureau of Radiation Protection and the NRC requested the Department of Energy to send a team from Brookhaven National Laboratory to assist in monitoring environmental radiation.53/

About this time, Mayor Robert Reid of Middletown telephoned Met Ed's home office in Reading. He was assured, he later told the Commission, that no radioactive particles had escaped and no one was injured.

> I felt relieved and relaxed; I said, "There's no problem." Twenty seconds later I walked out of my office and got in my car and turned the radio on and the announcer told me, over the radio, that there were radioactive particles released. Now, I said, "Gee whiz, what's going on here?" At 4:00 in the afternoon the same day the same man called me at home and said, "Mayor Reid, I want to update our conversation that we had at 11:00 a.m." I said, "Are you going to tell me that [radioactive] particles were released?" He said, "Yes." I said, "I knew that 20 seconds after I spoke to you on the phone."54/

Throughout much of the morning, Pennsylvania's Lieutenant Governor William Scranton, III, focused his attention on Three Mile Island. Scranton was charged, among other things, with overseeing the state's emergency preparedness functions. He had planned a morning press conference on energy conservation, but when he finally faced reporters in Harrisburg, the subject was TMI-2.55/ In a brief opening statement, Scranton said:

> The Metropolitan Edison Company has informed us that there has been an incident at Three Mile Island, Unit-2. Everything is under control. There is and was no danger to public health and safety. . . . There was a small release of radiation to the environment. All safety equipment functioned properly. Metropolitan Edison has been monitoring the air in the vicinity of the plant constantly since the incident. No increase in normal radiation levels has been detected 56/

During the questioning by reporters, however, William Dornsife of the state's Bureau of Radiation Protection, who was there at Scranton's invitation, said Met Ed employees had "detected a small amount of radioactive iodine. . . ." Dornsife had learned of the iodine reading (later found to be in error) just before the press conference began and had not had time to tell Scranton. Dornsife dismissed any threat to human health from the amount of radioactive iodine reported in Goldsboro.57/

Shortly after the press conference, a reporter told Scranton that Met Ed in Reading denied any off-site radiation. While some company executives were acknowledging radiation readings off the Island, low-level public relations officials at Met Ed's headquarters

ACCOUNT OF THE ACCIDENT

continued until noon to deny any off-site releases. It was an error in communications within Met Ed, one of several that would reduce the utility's credibility with public officials and the press. "This was the first contradictory bit of information that we received and it caused some disturbance," Scranton told the Commission in his testimony.58/

-0-0-0-

At Three Mile Island, the control room was crowded with operators and supervisors trying to bring the plant under control. They had failed in efforts to establish natural circulation cooling. This essentially means setting up a flow of water, without mechanical assistance, by heating water in the core and cooling it in the steam generators. This effort failed because the reactor coolant system was not filled with water and a gas bubble forming in the top of the reactor blocked this flow of water. At 11:38 a.m., operators began to decrease pressure in the reactor system. The pressurizer block valve was opened and high pressure injection cut sharply. This resulted again in a loss of coolant and an uncovering of the core. The depressurization attempt ended at 3:08 p.m.59/ The amount and duration of core uncovery during this period remains unknown.

About noon, three employees entered the auxiliary building and found radiation levels ranging from 50 millirems to 1,000 rems (one million millirems) an hour. Each of the three workers received an 800-millirem dose during the entry.60/ At 12:45 p.m., the Pennsylvania State Police closed Route 441 to traffic near Three Mile Island at the request of the state's Bureau of Radiation Protection. An hour later, the U.S. Department of Energy team began its first helicopter flight to monitor radiation levels. And at 1:50 p.m., a noise penetrated the TMI-2 control room; "a thud," as Gary Miller later characterized it.61/

That thud was the sound of a hydrogen explosion inside the containment building. It was heard in the control room; its force of 28 pounds per square inch was recorded on a computer strip chart there, which Met Ed's Michael Ross examined within a minute or two.62/ Yet Ross and others failed to realize the significance of the event. Not until late Thursday was that sudden and brief rise in pressure recognized as an explosion of hydrogen gas released from the reactor. The noise, said B&W's Leland Rogers in his deposition, was dismissed at the time as the slamming of a ventilation damper.63/ And the pressure spike on the strip chart, Ross explained to the Commission, "we kind of wrote it off . . . [as] possibly instrument malfunction. . . ."64/

Miller, Herbein, and Kunder left for Harrisburg soon afterwards for a 2:30 p.m. briefing with Lieutenant Governor Scranton on the events at Three Mile Island. At 2:27 p.m., radiation readings in Middletown ranged from 1 to 2 millirems per hour.65/

-0-0-0-

The influx of news media from outside the Harrisburg area began during the afternoon. The wire service reports of Associated Press

ACCOUNT OF THE ACCIDENT

TMI-2 control room several days after the start of the accident. The man in the foreground wearing a helmet is Charles Gallina, NRC inspector from Region I. The two men above Gallina are Craig Faust and William Zewe, both of whom were on duty when the accident began.

108

ACCOUNT OF THE ACCIDENT

and United Press International had alerted editors here and abroad to the accident. The heavy flow of newspaper and magazine reporters, television and radio correspondents, and photographers and camera crews would come later as the sense of concern about Three Mile Island grew. But at 4:30 p.m., when Scranton once more met the press, he found some strange faces among the familiar crew of correspondents who regularly covered Pennsylvania's Capitol.

Scranton had discussed the TMI situation with his own people and listened to Met Ed officials. "I wouldn't say that they [Met Ed] were exactly helpful, but they were not obstructive," he later testified. "I think they were defensive." Scranton was disturbed by, among other things, Herbein's comment during their 2:30 p.m. meeting that Herbein had not told reporters about some radiation releases during an earlier Met Ed press conference because "it didn't come up."66/ So Scranton was less assured about conditions at Three Mile Island when he issued his afternoon statement to the press:

> This situation is more complex than the company first led us to believe. We are taking more tests. And at this point, we believe there is still no danger to public health. Metropolitan Edison has given you and us conflicting information. We just concluded a meeting with company officials and hope this briefing will clear up most of your questions. There has been a release of radioactivity into the environment. The magnitude of the release is still being determined, but there is no evidence yet that it has resulted in the presence of dangerous levels. The company has informed us that from 11:00 a.m. until about 1:30 p.m., Three Mile Island discharged into the air, steam that contained detectable amounts of radiation. . . .67/

Scranton's statement inappropriately focused public attention on the steam emissions from TMI-2 as a source of radiation. In fact, they were not, since the water that flows inside the towers is in a closed loop and cannot mix with water containing radioactive materials unless there is a leak in the system.

Scranton went on to discuss potential health effects of the radiation releases:

> The levels that were detected were below any existing or proposed emergency action levels. But we are concerned because any increased exposure carries with it some increased health risks. The full impact on public health is being evaluated as environmental samples are analyzed. We are concerned most about radioactive iodine, which can accumulate in the thyroid, either through breathing or through drinking milk. Fortunately, we don't believe the risk is significant because most dairy cows are on stored feed at this time of year.68/

Many Americans learned about the accident at Three Mile Island from the evening newscasts of the television networks. Millions,

ACCOUNT OF THE ACCIDENT

Lieutenant Governor William Scranton (l) and Oran Henderson (r), director of the Pennsylvania Emergency Management Agency, at a March 28 press conference.

ACCOUNT OF THE ACCIDENT

for example, watched as Walter Cronkite led off the CBS Evening News:

> It was the first step in a nuclear nightmare; as far as we know at this hour, no worse than that. But a government official said that a breakdown in an atomic power plant in Pennsylvania today is probably the worst nuclear accident to date. . . .

At 7:30 p.m., Mayor Ken Myers of Goldsboro met with the borough council to discuss the accident and the borough's evacuation plan. Then Myers suggested he and the council members go door-to-door to talk with residents of the small community.

> Everyone listened to what we had to say. We mainly told them of what we had heard through the radio, TV, and even our own public relations and communications department in the basement of the York County court house. . . . Then we told them also of our evacuation plans in case the Governor would declare an emergency and that we would all have to leave. Of course, right away they gave us questions: "Well, what should we do? Do you think it's safe that we should stay or do you think we should go?" The ones that I talked to, I told them: "Use your own judgment. We dare not tell you to leave your homes."[69/]

ACCOUNT OF THE ACCIDENT

THURSDAY, MARCH 29

In retrospect, Thursday seemed a day of calm. A sense of betterment, if not well-being, was the spirit for much of the day. Radiation levels remained high at points within the auxiliary building, but off-site readings indicated no problems. The log book kept by the Dauphin County Office of Emergency Preparedness reflects this mood of a crisis passing:

5:45 a.m.	Called Pennsylvania Emergency Management Agency -- Blaisdale, reactor remains under control more stable than yesterday, not back to normal, monitoring continues by Met Ed, Radiological Health, and Nuclear Regulatory Commission.
7:55 a.m.	Pennsylvania Emergency Management Agency -- . . . no danger to public.
11:25 a.m.	Pennsylvania Emergency Management Agency advised situation same.
3:30 p.m.	. . . situation is improving.
6:12 p.m.	. . . no change -- not cold yet, continues to improve, slow rate, off-site release controlled.
7:00-9:00 p.m.	. . . Pennsylvania Emergency Management says Island getting better.
9:55 p.m.	. . . no real measureable reading off-site -- no health risk off-site, no emergency, bringing reactor to cold shut down. . . . 70/

Radiation monitoring continued. Midmorning readings showed 5 to 10 millirems an hour on-site and 1 to 3 millirems per hour across the Susquehanna River to the west. No radioactive iodine was detected in the air. The U.S. Food and Drug Administration began monitoring food, milk, and water in the area for radiation contamination. 71/

ACCOUNT OF THE ACCIDENT

Sen. Gary Hart (l), Sen. Alan Simpson (c), and Rep. Allen Ertel (r) arrive at TMI for a briefing on the accident, March 29.

ACCOUNT OF THE ACCIDENT

Thursday was a day of questioning. NRC Chairman Joseph Hendrie and several key aides journeyed to Capitol Hill to brief the House Subcommittee on Energy and the Environment and other members of Congress on the accident. Lieutenant Governor Scranton spent several hours in the early afternoon at Three Mile Island, touring the TMI-2 control room and auxiliary building, wearing a radiation suit and respirator during part of his inspection. That same afternoon, Met Ed officials and NRC inspectors briefed several visiting members of Congress, including Rep. Allen Ertel (D-Pa.), whose district includes Three Mile Island, and Sen. John Heinz (R-Pa.). Later in the day, a second Congressional delegation that included Sen. Richard Schweiker (R-Pa.) and Rep. William Goodling (R-Pa.), whose district includes York, Adams, and Cumberland counties, received a briefing.

Thursday was also a day of disquieting discussions and discoveries. Thursday afternoon, a telephone conversation took place between two old acquaintances, Gordon MacLeod, Pennsylvania's Secretary of Health, and Anthony Robbins, director of the National Institute for Occupational Safety and Health. One important point of that conversation remains in dispute. MacLeod recalls that Robbins urged him to recommend an evacuation of people living around Three Mile Island.72/ Robbins denies discussing or suggesting such an evacuation.73

Up to this point, MacLeod -- who had taken office only 12 days before the accident -- had offered no recommendations since his department had no direct responsibility for radiological health matters. Now, however, he arranged a conference telephone call with Oran Henderson, director of the Pennsylvania Emergency Management Agency; Thomas Gerusky, director of the Bureau of Radiation Protection; and John Pierce, an aide to Lieutenant Governor Scranton. MacLeod told them Robbins had strongly recommended evacuation. The others rejected the idea, although they agreed it should be reconsidered if conditions proved worse than they appeared at TMI-2. MacLeod then asked if it might be wise to have pregnant women and children under age 2 leave the area around the nuclear plant. This, too, was rejected Thursday afternoon.74/

At 2:10 p.m., a helicopter over TMI-2 detected a brief burst of radiation that measured 3,000 millirems per hour 15 feet above the plant's vent. This information was relayed to NRC headquarters, where it created no great concern.

But another release that afternoon, one within NRC limits for radiation releases, did cause considerable consternation. Soon after the accident began Wednesday, Met Ed stopped discharging wastewater from such sources as toilets, showers, laundry facilities, and leakage in the turbine and control and service buildings into the Susquehanna River. Normally, this water contains little or no radioactivity, but as a result of the accident, some radioactive gases had contaminated it. The radiation levels, however, were within the limits set by the NRC. By Thursday afternoon, nearly 400,000 gallons of this slightly radioactive water had accumulated and the tanks were now close to overflowing. Two NRC officials -- Charles Gallina on-site and George Smith at the Region I office --

ACCOUNT OF THE ACCIDENT

told Met Ed they had no objections to releasing the water so long as it was within NRC specifications. Met Ed notified the Bureau of Radiation Protection and began dumping the wastewater. No communities downstream from the plant were informed, nor was the press.75/

When NRC Chairman Hendrie learned of the release, he ordered it stopped. Hendrie did not know the water's source, and he was concerned about the impact on the public of the release of any radiation, no matter how slight.76/ Some 40,000 gallons had entered the river when the dumping ceased around 6:00 p.m. Both NRC officials on-site and the Governor's aides realized that authorizing release of the wastewater would be unpopular, and neither was eager to do so. Yet the tanks still were close to overflowing. After hours of discussion, agreement was reached on the wording of a press release that the state's Department of Environmental Resources issued, which said DER "reluctantly agrees that the action must be taken." Release of the wastewater resumed shortly after midnight.77/

Late Thursday afternoon, Governor Thornburgh had held a press conference. At it, the NRC's Charles Gallina told reporters the danger was over for people off the Island. Thornburgh distrusted the statement at the time, and events soon confirmed his suspicion. At 6:30 p.m., Gallina and James Higgins, an NRC reactor inspector, received the results of an analysis of the reactor's coolant water. It showed that core damage was far more substantial than either had anticipated. At 10:00 p.m., Higgins telephoned the Governor's office with the new information and indicated that a greater possibility of radiation releases existed. Nothing had changed inside the plant, only NRC's awareness of the seriousness of the damage. Yet Higgins' call foretold events only hours away.78/

ACCOUNT OF THE ACCIDENT

FRIDAY, MARCH 30

The TMI-2 reactor has a means of removing water from the reactor coolant system, called the let-down system, and one for adding water, called the make-up system. Piping from both runs through the TMI-2 auxiliary building, and NRC officials suspected that leaks in these two systems explained the sporadic, uncontrolled releases of radioactivity. They were also concerned about levels in the make-up tank and the two waste gas decay tanks inside the auxiliary building. Water from the let-down system flows into the make-up tank. In that tank, gases dissolved in the reactor's cooling water at high pressure are released because the tank's pressure is lower, much as the gas bubbles in a pressurized carbonated beverage appear when the bottle is opened. These gases, under normal circumstances, are compressed and stored in the waste gas decay tanks. NRC officials worried that if the waste gas decay tanks filled to capacity, relief valves would open, allowing a continuing escape of radiation into the environment. 79/ That concern and what Commission Chairman Kemeny would later call a "horrible coincidence" 80/ resulted in a morning of confusion, contradictory evacuation recommendations, and eventually an evacuation advisory from Governor Richard Thornburgh.

About halfway through his midnight-to-noon shift on Friday, James Floyd, TMI-2's supervisor of operations, decided to transfer radioactive gases from the make-up tank to a waste gas decay tank. Floyd knew this would release radiation because of leaks in the system, but he considered the transfer necessary. The pressure in the make-up tank was so high that water that normally flowed into it for transfer to the reactor coolant system could not enter the tank. Floyd, without checking with other TMI and Met Ed officials, ordered the transfer to begin at 7:10 a.m. to reduce the tank's pressure. This controlled release allowed radioactive material to escape into the auxiliary building and then into the air outside. Thirty-four minutes later, Floyd requested a helicopter be sent to take radiation measurements. The chopper reported readings of 1,000 millirems per hour at 7:56 a.m. and 1,200 millirems per hour at 8:01 a.m., 130 feet above the TMI-2 vent stack. 81/

116

ACCOUNT OF THE ACCIDENT

A helicopter taking air samples over the containment building.

ACCOUNT OF THE ACCIDENT

-0-0-0-

At NRC headquarters, Lake Barrett, a section leader in the environmental evaluation branch, was concerned about the waste gas decay tank level. The previous evening, he had helped calculate "a hypothetical release rate" for the radiation that would escape if the tank's relief valves opened. Shortly before 9:00 a.m., Barrett was told of a report from Three Mile Island that the waste gas decay tanks had filled. He was asked to brief senior NRC staff officials on the significance of this. The group included Lee Gossick, executive director for operations; John Davis, then acting director of Inspection and Enforcement; Harold Denton, director of Nuclear Reactor Regulation; Victor Stello, Jr., then director of the Office of Operating Reactors; and Harold Collins, assistant director for emergency preparedness in the Office of State Programs. During the briefing, Barrett was asked what the release rate would mean in terms of an off-site dose. He did a quick calculation and came up with a figure: 1,200 millirems per hour at ground level. Almost at that moment, someone in the room reported a reading of 1,200 millirems per hour had been detected at Three Mile Island. By coincidence, the reading from TMI was identical to the number calculated by Barrett. "It was the exact same number, and it was within maybe 10 or 15 seconds from my first 1,200 millirems per hour prediction," Barrett told the Commission.82/

The result was instant concern among the NRC officials; "an atmosphere of significant apprehension," as Collins described it in his testimony. Communications between the NRC headquarters and Three Mile Island had been less than satisfactory from the beginning. "I think there was uncertainty in the operations center as to precisely what was going on at the facility and the question was being raised in the minds of many as to whether or not those people up there would do the right thing at the right time, if it had to be done," Collins testified. NRC officials proceeded without confirming the reading and without knowing whether the 1,200 millirem per hour reading was on- or off-site, whether it was taken at ground level or from a helicopter, or what its source was. They would later learn that the radiation released did not come from the waste gas decay tanks. The report that these tanks had filled was in error.83/

After some discussion, Harold Denton directed Collins to notify Pennsylvania authorities that senior NRC officials recommended the Governor order an evacuation. Collins telephoned Oran Henderson, director of the Pennsylvania Emergency Management Agency, and, apparently selecting the distance on his own, recommended an evacuation of people as far as 10 miles downwind from Three Mile Island. Henderson telephoned Lieutenant Governor Scranton, who promised to call the Governor. A Henderson aide also notified Thomas Gerusky, director of the Bureau of Radiation Protection, of the evacuation recommendation. Gerusky knew of the 1,200 millirem reading. A telephone call to an NRC official at the plant reinforced Gerusky's belief that an evacuation was unnecessary. He tried to telephone Governor Thornburgh, found the lines busy, and went to the Governor's office to argue personally against an evacuation.84/

ACCOUNT OF THE ACCIDENT

Kevin Molloy, director of emergency preparedness for Dauphin County, had received a call from Met Ed's James Floyd at 8:34 a.m., alerting him to the radiation release. Twenty minutes later, the Pennsylvania Emergency Management Agency notified Molloy of an on-site emergency and an increase in radiation, but Molloy was told that no evacuation was needed. Then at 9:25 a.m., Henderson called Molloy and told him to expect an official evacuation order in 5 minutes; the emergency preparedness offices in York and Lancaster counties received similar alerts. Molloy began his preparations. He notified all fire departments within 10 miles of the stricken plant, and broadcast a warning over radio station WHP that an evacuation might be called.85/

At Three Mile Island, NRC's Charles Gallina was confronted by a visibly upset Met Ed employee shortly after Molloy's broadcast. "As the best I can remember, he said, 'What the hell are you fellows doing? My wife just heard the NRC recommended evacuation,'" Gallina told the Commission. Gallina checked radiation readings on- and off-site and talked with an NRC reactor inspector, who said "things were getting better." Then Gallina telephoned NRC officials at Region I and at Bethesda headquarters in an attempt "to call back that evacuation notice."86/

Shortly after 10:00 a.m., Governor Thornburgh talked by telephone with Joseph Hendrie. The NRC chairman assured the Governor that no evacuation was needed. Still, Hendrie had a suggestion: that Thornburgh urge everyone within 5 miles downwind of the plant to stay indoors for the next half-hour. The Governor agreed and later that morning issued an advisory that all persons within 10 miles of the plant stay inside. During this conversation, Thornburgh asked Hendrie to send a single expert to Three Mile Island upon whom the Governor could rely for technical information and advice.87/

About an hour later, Thornburgh received a telephone call from President Carter, who had just talked with Hendrie. The President said that he would send the expert the Governor wanted. That expert would be Harold Denton. The President also promised that a special communications system would be set up to link Three Mile Island, the Governor's office, the White House, and the NRC.88/

Thornburgh convened a meeting of key aides to discuss conditions at Three Mile Island. During this meeting, at about 11:40 a.m., Hendrie again called the Governor. As Gerusky recalls the conversation that took place over a speaker phone, the NRC chairman apologized for the NRC staff error in recommending evacuation. Just before the call, Emmett Welch, an aide to Gordon MacLeod, had renewed the Secretary of Health's recommendation that pregnant women and children under age 2 be evacuated. Thornburgh told Hendrie of this. Gerusky recalls this response from Hendrie: "If my wife were pregnant and I had small children in the area, I would get them out because we don't know what is going to happen."89/ After the call, Thornburgh decided to recommend that pregnant women and preschool children leave the region within a 5-mile radius of Three Mile Island and to close all schools within that area. He issued his advisory shortly after 12:30 p.m.

ACCOUNT OF THE ACCIDENT

Thornburgh was conscious throughout the accident that an evacuation might be necessary, and this weighed upon him. He later shared some of his concerns in testimony before the Commission:

> There are known risks, I was told, in an evacuation. The movement of elderly persons, people in intensive care units, babies in incubators, the simple traffic on the highways that results from even the best of an orderly evacuation, are going to exert a toll in lives and injuries. Moreover, this type of evacuation had never been carried out before on the face of this earth, and it is an evacuation that was quite different in kind and quality than one undertaken in time of flood or hurricane or tornado. . . . When you talk about evacuating people within a 5-mile radius of the site of a nuclear reactor, you must recognize that that will have 10-mile consequences, 20-mile consequences, 100-mile consequences, as we heard during the course of this event. This is to say, it is an event that people are not able to see, to hear, to taste, to smell. . . . 90/

-0-0-0-

Relations between reporters and Met Ed officials had deteriorated over several days. Many reporters suspected the company of providing them with erroneous information at best, or of outright lying. When John Herbein arrived at 11:00 a.m. Friday to brief reporters gathered at the American Legion Hall in Middletown, the situation worsened. The press corps knew that the radioactivity released earlier had been reported at 1,200 millirems per hour; Herbein did not. He opened his remarks by stating that the release had been measured at around 300 to 350 millirems per hour by an aircraft flying over the Island. The question-and-answer period that followed focused on the radiation reading -- "I hadn't heard the number 1,200," Herbein protested during the news conference -- whether the release was controlled or uncontrolled, and the previous dumping of radioactive wastewater. At one point Herbein said, "I don't know why we need to . . . tell you each and every thing that we do specifically. . . ." 91/ It was that remark that essentially eliminated any credibility Herbein and Met Ed had left with the press. 92/

The next day, Jack Watson, a senior White House aide, would telephone Herman Dieckamp, president of Met Ed's parent company, to express his concern that the many conflicting statements about TMI-2 reported by the news media were increasing public anxiety. Watson would suggest that Denton alone brief reporters on the technical aspects of the accident and Dieckamp would agree. 93/

-0-0-0-

The radiation release, Molloy's announcement of a probable evacuation, and finally the Governor's advisory brought concern and even fear to many residents. Some people had already left, quietly evacuating on their own; others now departed. "On March 29 of this year, my wife and I joyously brought home our second daughter from the hospital; she was just 6 days old," V.T. Smith told the Commission.

ACCOUNT OF THE ACCIDENT

Governor Richard Thornburgh testifying before the Commission.

ACCOUNT OF THE ACCIDENT

Map of the TMI area showing 5-, 10-, and 20-mile evacuation zones.

ACCOUNT OF THE ACCIDENT

"On the morning of the 30th, all hell broke loose and we left for Delaware to stay with relatives."94/ By Saturday evening, a Goldsboro councilman estimated 90 percent of his community's residents had left.95/

Schools closed after the Governor's advisory. Pennsylvania State University called off classes for a week at its Middletown campus. Friday afternoon, " . . . still having heard nothing from Three Mile Island," Harrisburg Mayor Paul Doutrich drove with his deputy public works director to the TMI Observation Center overlooking the nuclear facility. There they talked for an hour with Met Ed President Creitz and Vice President Herbein. "Oddly enough, one of the things that impressed me the most and gave me the most feeling of confidence that things were all right was that everybody in that area, all the employees, the president and so forth, were walking around in their shirt sleeves, bare-headed," Doutrich told the Commission. "I saw not one indication of nuclear protection."96/

Friday, Saturday, and Sunday were hectic days in the emergency preparedness offices of the counties close to Three Mile Island. Officials labored first to prepare 10-mile evacuation plans and then ones covering areas out to 20 miles from the plant. The Pennsylvania Emergency Management Agency recommended Friday morning that 10-mile plans be readied. The three counties closest to the nuclear plant already had plans to evacuate their residents -- a total of about 25,000 living within 5 miles of the Island. A 10-mile evacuation had never been contemplated. For Kevin Molloy in Dauphin County, extending the evacuation zone meant the involvement of several hospitals -- something he had not confronted earlier. There were no hospitals within 5 miles. Late Friday night, PEMA told county officials to develop 20-mile plans. Suddenly, six counties were involved in planning for the evacuation of 650,000 people, 13 hospitals, and a prison.97/

Friday was also the day the nuclear industry became deeply involved in the accident. After the radiation release that morning, GPU President Dieckamp set about assembling an industry team to advise him in managing the emergency. Dieckamp and an aide talked with industry leaders around the country, outlining the skills and knowledge needed at TMI-2. By late Saturday afternoon, the first members of the Industry Advisory Group had arrived. They met with Dieckamp, identified the tasks that needed immediate attention, and decided who would work on each.98/

-0-0-0-

Harold Denton arrived on site about 2:00 p.m. Friday, bringing with him a cadre of a dozen or so experts from NRC headquarters. Earlier in the day, NRC had learned of the hydrogen burn or explosion that flashed through the containment building Wednesday afternoon. The NRC staff already knew that some form of gas bubble existed within the reactor system. Now it became obvious that the bubble, an estimated 1,000 cubic feet of gases, contained hydrogen. And as Denton would later recall in his deposition, the question arose whether there was a potential for a hydrogen explosion. Throughout

ACCOUNT OF THE ACCIDENT

Friday, Denton operated on estimates provided him before he left Bethesda, which indicated that the bubble could not self-ignite for 5 to 8 days. Denton focused his immediate attention on finding ways to eliminate the bubble.99/

At about 8:30 p.m. Friday, Denton briefed Governor Thornburgh in person for the first time. Fuel damage was extensive; the bubble posed a problem in cooling the core; no immediate evacuation was necessary, Denton said. Then the two men held their first joint press conference. The Governor reiterated that no evacuation was needed, lifted his advisory that people living within 10 miles of Three Mile Island stay indoors, but continued his recommendation that pregnant women and preschool children remain more than 5 miles from the plant.100/

-0-0-0-

Shortly after 4:00 p.m., Jack Watson, President Carter's assistant for intergovernmental affairs, called Jay Waldman, Governor Thornburgh's executive assistant. The two disagree about the substance of that call. In an interview with the Commission staff, Waldman said Watson asked that the Governor not request President Carter to declare a state of emergency or disaster:

> He said that it was their belief that that would generate unnecessary panic, that the mere statement that the President has declared this area an emergency and disaster area would trigger a substantial panic; and he assured me that we were getting every type and level of federal assistance that we would get if there had been a declaration. I told him that I would have to have his word on that, an absolute assurance, and that if that were true, I would go to the Governor with his request that we not formally ask for a declaration.101/

Watson and his assistant, Eugene Eidenberg, both said in their Commission depositions that the White House never asked Governor Thornburgh not to request such a declaration.102/ Whatever was said in that Friday conversation, the Governor made no request to the President for an emergency declaration. State officials later expressed satisfaction with the assistance provided by the federal government during the accident and immediately after. They were less satisfied, however, in August with the degree of assistance and cooperation they were receiving from federal agencies.103/

-0-0-0-

Officials of the U.S. Department of Health, Education, and Welfare (HEW) had become concerned about the possible release of radioactive iodine at Three Mile Island and began Friday to search for potassium iodide -- a drug capable of preventing radioactive iodine from lodging in the thyroid. The thyroid absorbs potassium iodide to a level where the gland can hold no more. Thus, if a person is exposed to radioactive iodine after receiving a sufficient quantity of potassium iodide, the thyroid is saturated and cannot absorb the additional iodine with its potentially damaging radiation. At the time of the TMI-2 accident, however, no pharmaceutical or chemical company was marketing medical-grade potassium iodide in the quantities needed.104/

Saturday morning, shortly after 3:00 a.m., the Mallinckrodt Chemical Company agreed to provide HEW with approximately a quarter million one-ounce bottles of the drug. Mallinckrodt in St. Louis, working with Parke-Davis in Detroit and a bottle-dropper manufacturer in New Jersey, began an around-the-clock effort. The first shipment of potassium iodide reached Harrisburg about 1:30 a.m. Sunday. By the time the last shipment arrived on Wednesday, April 4, the supply totalled 237,013 bottles.105/

ACCOUNT OF THE ACCIDENT

SATURDAY, MARCH 31

The great concern about a potential hydrogen explosion inside the TMI-2 reactor came with the weekend. That it was a groundless fear, an unfortunate error, never penetrated the public consciousness afterward, partly because the NRC made no effort to inform the public it had erred. 106/

Around 9:30 p.m. Friday night, the NRC chairman asked Roger Mattson to explore the rate at which oxygen was being generated inside the TMI-2 reactor system and the risk of a hydrogen explosion. "He said he had done calculations," Mattson said in his deposition. "He was concerned with the answers." 107/ Mattson is director of the Division of Systems Safety within the Office of Nuclear Reactor Regulation (NRR), which is headed by Denton, and had spent part of Thursday and Friday working on how to remove a gas bubble from the reactor. Following Denton's departure for TMI, Mattson served variously as NRR's representative or deputy representative at the Incident Response Center.

Hydrogen had been produced in the reactor as a result of a high-temperature reaction that occurred between hot steam and the zirconium cladding of the fuel rods. For this hydrogen to explode or burn -- a less dangerous possibility -- enough oxygen would have to enter the system to form an explosive mixture. There were fears this would happen as the result of radiolysis. In this process, radiation breaks apart water molecules, which contain hydrogen and oxygen.

Two NRC teams worked throughout the weekend on the problem, and both sought help from laboratories and scientists outside the NRC. One group addressed the rate at which radiolysis would generate oxygen at TMI-2. The second analyzed the potential for hydrogen combustion. Robert Budnitz of the NRC also asked experts about possible chemicals that might remove the hydrogen.

At noon, Hendrie talked by telephone with Denton and expressed his concern that oxygen freed by radiolysis was building up in the reactor. Earlier, Hendrie had told Victor Stello, Jr., Denton's second-in-command at TMI, the same thing. The NRC chairman told

ACCOUNT OF THE ACCIDENT

NRC Chairman Joseph Hendrie testifying at a Commission hearing.

ACCOUNT OF THE ACCIDENT

Harold Denton, director of NRC's Office of Nuclear Reactor Regulation, and Governor Thornburgh at a March 31 news conference in Harrisburg.

ACCOUNT OF THE ACCIDENT

Denton that Governor Thornburgh should be made aware of the potential danger. Denton promised to speak with Thornburgh.

Shortly after 1:00 p.m., Mattson got some preliminary answers regarding the potential for a hydrogen explosion. An hour later, Mattson got more replies. "I had an estimate there was oxygen being generated, from four independent sources, all with known credentials in this field," he said in his deposition. "The estimate of how much oxygen varied, but all estimates said there was considerable time, a matter of several days, before there was a potential combustible mixture in the reactor coolant system."108/

At a Commission hearing, Mattson later admitted in response to questions from Commissioner Pigford that the NRC could have determined from the information available at that time that no excess oxygen was being generated and there was no real danger of explosion.109/ But when Mattson met with the NRC commissioners at 3:27 p.m. on Saturday, "the bottom line of that conversation . . . was there were several days required to reach the flammability limit, although there was oxygen being generated," Mattson recalled in his deposition. "And I expressed confidence that we were not underestimating the reactor coolant system explosion potential; that is, the estimate of 2 to 3 days before reaching the flammability limit was a conservative estimate." By Saturday night, however, Mattson would be told by his consultants that their calculations indicated that the oxygen percentage of the bubble was on the threshold of the flammability limit.110/

Around 6:45 p.m., Mattson talked with Vincent Noonan, the man within NRC most knowledgeable about what might result from an explosion inside a reactor. One NRC consultant had predicted that a hydrogen blast would produce pressures of 20,000 pounds per square inch within the TMI-2 reactor. B&W, designer of the reactor, however, had considered the dampening effects of water vapor on an explosion and those of an enriched hydrogen environment and calculated a total pressure of 3,000 to 4,000 psi. That was encouraging. Previous analyses indicated the reactor coolant system of a TMI-2 reactor could withstand blast pressures of that magnitude.

Late Saturday evening, James Taylor of B&W reiterated another B&W engineer's conclusion first relayed to the NRC Thursday night -- that no excess oxygen was being generated. That information, Mattson stated in his deposition, never reached him.111/

Saturday at 2:45 p.m., Hendrie met with reporters in Bethesda. He said then that a precautionary evacuation out to 10 or 20 miles from the Island might be necessary if engineers attempted to force the bubble out of the reactor. NRC had concluded such an attempt might cause further damage to the core, Hendrie said, and it might touch off an explosion of the bubble.

Stan Benjamin, a reporter with the Washington bureau of the Associated Press, followed up Hendrie's press conference by interviewing two NRC officials: Edson Case, Denton's deputy in the Office of Nuclear Reactor Regulation, and Frank Ingram, a public information spokesman. From them, and an NRC source he refused to name, Benjamin learned of the concern within the Incident Response

ACCOUNT OF THE ACCIDENT

Center that the bubble could become a potentially explosive mixture within a matter of days, perhaps as few as two. Benjamin checked his story with Case and Ingram, reading much of it to them word by word, before releasing the article. Case and Ingram agreed it was accurate. The report -- first transmitted as an editor's note at 8:23 p.m. -- was the first notice to the public that some NRC officials feared the bubble might possibly explode spontaneously.112/

Denton had been briefed throughout Saturday afternoon and evening by Hendrie and NRC officials in Bethesda on the oxygen estimates and the potential for a burn or explosion. But he learned of the AP story only a short time before he joined Governor Thornburgh and Lieutenant Governor Scranton for a late evening press conference in Harrisburg. The Governor assured reporters that "there is no imminent catastrophic event foreseeable at the Three Mile Island facility." Denton, too, said: "There is not a combustible mixture in the containment or in the reactor vessel. And there is no near-term danger at all." Denton also tried to deflate the impression, voiced by several reporters, that contradictions existed between himself and his colleagues at NRC headquarters. "No, there is no disagreement. I guess it is the way things get presented," he said.113/

But there was disagreement, and Denton wanted it resolved. President Carter had announced earlier in the evening he would visit TMI the following day. Denton told Stello to explore the oxygen-hydrogen issue further with outside experts. Stello realized the concern in Washington. He had received a telephone call shortly after 9:00 p.m. from Eugene Eidenberg, a Presidential aide, inquiring about the AP story. Stello told the White House that he did not share the concern felt at NRC headquarters.

-0-0-0-

Saturday, as the NRC wrestled with managing the accident and the envisioned danger of the hydrogen bubble, officials of the Department of Health, Education, and Welfare struggled with their own concerns. That morning, senior HEW health officials gathered and continued the previous day's discussion of the possibility of an evacuation; for the first time, they debated how large an area should be evacuated. But the discussions led ultimately to a recommendation to consider immediate evacuation if the NRC could not provide assurances that the reactor was cooling safely. Joseph Califano, HEW Secretary, summarized the group's views in a memorandum to Jack Watson of the President's staff.

Later in the day, HEW health officials attended an interagency meeting at the White House, convened by Watson, and repeated the HEW recommendation to consider evacuation. Richard Cotton, a key Califano aide, raised another Califano recommendation that NRC officials consult with HEW and Environmental Protection Agency experts regarding the potential health effects of the efforts to control TMI-2's reactor. Cotton persisted after the meeting, and on Sunday and Tuesday HEW officials were briefed by the NRC. These briefings, however, were always informational; there was no NRC effort to seek HEW's advice.114/

ACCOUNT OF THE ACCIDENT

SUNDAY, APRIL 1

Throughout Saturday night and the early hours of Sunday, county emergency preparedness offices were deluged with telephone calls from citizens concerned by the conflicting reports about the hydrogen bubble. But the flow of useful information from the state to the local level had essentially ceased after Denton's arrival. The Governor's office focused attention on the federal effort -- Denton and officials from several U.S. emergency agencies. Oran Henderson, director of the Pennsylvania Emergency Management Agency, was no longer invited to the Governor's briefings and press conferences, and he did not attend after Friday night. Thus PEMA -- although it continued to receive status reports from the Bureau of Radiation Protection -- was isolated from information wanted at the local level.

In Dauphin County, frustration ran high. Shortly before midnight on Saturday, State Sen. George Gekas called the Governor in an attempt to obtain accurate information. Gekas was told the Governor was too busy to talk. Then Gekas called Scranton, and got the same response. At that point, Gekas told a Scranton aide that unless more cooperation and information were forthcoming, Dauphin County would order an evacuation at 9:00 a.m. Sunday. Scranton called the county's emergency center at 2:00 a.m. and agreed to meet officials there later in the morning. The Lieutenant Governor arrived at 10:00 a.m., preceded by Henderson, who complained of his own inability to obtain information. Scranton listened to Molloy and his colleagues. "I think he was just totally shocked by what was transpiring at our level; how busy we were; how much work we were doing; how complicated it was," Molloy said in his deposition.115/

-0-0-0-

Sunday, Mattson and several other NRC staffers met with NRC Commissioners Hendrie, Victor Gilinsky, and Richard Kennedy. Their purpose was to reach a judgment, based on the estimates and information available, about the true potential for a hydrogen explosion inside the reactor. According to Mattson's deposition, the group agreed:

ACCOUNT OF THE ACCIDENT

President Jimmy Carter touring the TMI-2 control room with (l to r) Harold Denton, Governor Thornburgh, and James Floyd, supervisor of TMI-2 operations, on April 1.

ACCOUNT OF THE ACCIDENT

> 5 percent oxygen was a realistic flammability limit, 11 percent oxygen was a realistic detonation limit, that there could be no spontaneous combustion below 900°F, that the oxygen production rate was approximately one percent per day, and that the present oxygen concentration in the bubble was 5 percent. 116/

After the meeting, Hendrie and Mattson drove to TMI to meet with Denton.

> Stello talked with Denton Sunday morning and outlined his arguments against any danger of a hydrogen explosion inside the reactor. Pressurized water reactors, the type used at TMI-2, normally operate with some free hydrogen in the reactor coolant. This hydrogen joins with the oxygen freed by radiolysis to form another water molecule, which prevents the build-up of oxygen to a quantity that would allow an explosion to take place. Stello told Denton that the process was the same now, and there was no danger of explosion.

> Hendrie and Mattson met with Denton and Stello in a hangar at Harrisburg International Airport minutes before President Carter's 1:00 p.m. arrival. Mattson and Stello had not talked to each other since Friday morning. Mattson outlined the conclusions reached at NRC headquarters about the bubble and the reasoning behind them. In an interview with the Commission staff, Mattson described what happened next:

>> And Stello tells me I am crazy, that he doesn't believe it, that he thinks we've made an error in the rate of calculation . . . . Stello says we're nuts and poor Harold is there, he's got to meet with the President in 5 minutes and tell it like it is. And here he is. His two experts are not together. One comes armed to the teeth with all these national laboratories and Navy reactor people and high faluting PhDs around the country, saying this is what it is and this is his best summary. And his other [the operating reactors division] director, saying, "I don't believe it. I can't prove it yet, but I don't believe it. I think it's wrong."117/

> Upon the President's arrival, Denton briefed the Chief Executive on the status of the plant and the uncertainty regarding its infamous bubble.

> The President was driven to TMI, put on protective yellow plastic shoecovers, and toured the facility with Mrs. Carter, Governor Thornburgh, and Denton. Stello, Hendrie, and Mattson went to the temporary NRC offices. During the afternoon, experts -- including those at Westinghouse and General Electric -- were canvassed by phone. "By three o'clock, we're convinced we've got it," Mattson said in his interview. "It's not going to go boom."118/

> NRC scientists in Bethesda eventually reached the same conclusion, but later in the day. Shortly before 4:00 p.m., NRC Commissioners Richard Kennedy, Peter Bradford, and John Ahearne met. They expressed concern over the differing estimates presented by the NRC staff and decided there might be a need to consider evacuation.

ACCOUNT OF THE ACCIDENT

Kennedy telephoned Hendrie at TMI and told him the three NRC Commissioners thought Governor Thornburgh should advise a precautionary evacuation within 2 miles of the plant, unless experts on-site had better technical information than that available in Bethesda.119/ Hendrie assured Kennedy that the free hydrogen inside the reactor would capture any oxygen generated and that no problem existed.

In midafternoon, new measurements showed the large bubble in the reactor was diminishing. The gases still existed, but they were distributed throughout the system in smaller bubbles that made eliminating the predominantly hydrogen mixture easier. Why this occurred, no one knows. But it was not because of any intentional manipulation by Met Ed or NRC engineers.

By late Sunday afternoon, NRC -- which was responsible for the concern that the bubble might explode -- knew there was no danger of a blast and that the bubble appeared to be diminishing. It was good news, but good news unshared with the public. Throughout Sunday, the NRC made no announcement that it had erred in its calculations or that no threat of an explosion existed. Governor Thornburgh was not told of the NRC miscalculation either. Nor did the NRC reveal the bubble was disappearing that day, partly because the NRC experts themselves were not absolutely certain.

ACCOUNT OF THE ACCIDENT

MONDAY, APRIL 2

Monday morning Denton and Mattson met the press. George Troffer, a Met Ed official, had already told a reporter the bubble was essentially gone. Denton acknowledged a "dramatic decrease in bubble size," but cautioned that more sophisticated analyses were needed "to be sure that the equations that are used to calculate bubble size properly include all effects." As to the bubble's potential for explosion, Denton told reporters "the oxygen generation rate that I was assuming yesterday when I was reporting on the potential detonation inside the vessel is, it now appears, to have been too conservative." Throughout the press conference, Denton continued to refer to NRC's estimates as too conservative; he never stated outright that NRC had erred in its conclusion that the bubble was near the dangerous point.[120]/

According to Mattson, the tone of the press conference -- its vagueness and imprecision -- was decided upon at a meeting of NRC officials Monday morning.

> We wanted to go slow on saying it was good news. We wanted to say it is good news, do not panic, we think we have got it under control, things look better, but we did not want to firmly and finally conclude that there was no problem. We had to save some wiggle room in order to preserve credibility. That was our judgement."[121]/

ACCOUNT OF THE ACCIDENT

Harold Denton at an April 2 press conference. The bearded man behind Denton is Roger Mattson, director of NRC's Division of Systems Safety. Beside Denton is Joseph Fouchard, NRC's public affairs director.

136

ACCOUNT OF THE ACCIDENT

EPILOGUE

The accident at Three Mile Island did not end with the breaking up of the bubble, nor did the threat to the health and safety of the workers and the community suddenly disappear. A small bubble remained, gases still existed within TMI-2's cooling water, and the reactor itself was badly damaged. Periodic releases of low-level radiation continued, and some feared a major release of radioactive iodine-131 might yet occur. Schools remained closed. The Governor's recommendation that pregnant women and preschool children stay more than 5 miles from the plant continued.

Saturday, March 31, the Department of Health, Education, and Welfare had arranged for the rapid manufacture of nearly a quarter million bottles of potassium iodide.122/ That same day, the Pennsylvania Bureau of Radiation Protection -- which had originally accepted HEW's offer to obtain the drug -- transferred responsibility for handling the radioactive iodine blocker to the state's Department of Health. Gordon MacLeod, who headed the health department, put the drug shipments in a warehouse as they began arriving Sunday. During the weekend, Thomas Gerusky, director of the Bureau of Radiation Protection, requested that his people at TMI be issued potassium iodide; Gerusky wanted BRP personnel to have the thyroid-blocking agent available should a release of radioactive iodine occur. MacLeod refused. He argued that if the public learned that any of the drug had been issued, a demand for its public distribution would result.

MacLeod had the backing of the Governor's office and Harold Denton in his decision not to issue the potassium iodide. The decision did not find agreement in Washington, however. On Monday, Jack Watson asked HEW to prepare recommendations for the drug's distribution and use. These were developed by a group headed by Donald Frederickson, director of the National Institutes of Health. The recommendation included: administering potassium iodide immediately to all workers on the Island; providing the drug to all people who would have less than 30 minutes' warning of a radioactive iodine release (roughly those within 10 miles of the plant); and that local authorities assess these recommendations in light of their first-hand knowledge of the situation.

ACCOUNT OF THE ACCIDENT

Governor Thornburgh received the recommendations in a White House letter on Tuesday, although some Pennsylvania officials had learned of them Monday. MacLeod strongly opposed distributing the drug to the public. Among his reasons: radioiodine levels were far below what was indicated for protective action, and the likelihood of a high-level release from TMI-2 was diminishing; distributing the drug would increase public anxiety and people might take it without being told to do so; and the possibility of adverse side-effects presented a potential public health problem in itself. MacLeod chose not to accept the federal recommendations. The potassium iodide remained in a warehouse under armed guard throughout the emergency. In midsummer, the FDA moved the drug to Little Rock, Arkansas, for storage.

-0-0-0-

Tuesday, April 3, General Public Utilities, Met Ed's parent company, established its TMI-2 recovery organization to oversee and direct the long process of cleaning up TMI-2. Robert Arnold, a vice president of another subsidiary, the GPU Service Corporation, was named to head the recovery operation.123/

Wednesday, April 4, schools outside the 5-mile area surrounding TMI reopened. All curfews were lifted. But schools within 5 miles of the Island remained closed and the Governor's advisory remained in effect for pregnant women and preschool children.

Some sense of normalcy was gradually returning to the TMI area. Governor Thornburgh asked Denton repeatedly if the advisory could be lifted, allowing pregnant women and preschool children to return home. But the NRC wanted some specific event as a symbol to announce the crisis had ended. At first, the NRC looked to reaching "cold shutdown" -- the point at which the temperature of TMI-2's reactor coolant fell below the boiling point of water. When it became obvious that cold shutdown was days away, agreement was reached between Pennsylvania's Bureau of Radiation Protection and the NRC on ending the advisory. On Saturday, April 7, Kevin Molloy, at the request of the Governor's office, read a press release announcing the closing of the evacuation shelter at the Hershey Park Arena. Not until 2 days later, however, did Governor Thornburgh officially withdraw the advisory.124/

-0-0-0-

The accident at TMI did not end with cold shutdown, nor will it end for some time. More than a million gallons of radioactive water remain inside the containment building or stored in auxiliary building tanks. The containment building also holds radioactive gases and the badly damaged and highly radioactive reactor core. Radioactive elements contaminate the walls, floors, and equipment of several buildings. Ahead lies a decontamination effort unprecedented in the history of the nation's nuclear power industry -- a cleanup whose total cost is estimated at $80 to $200 million and which will take several years to complete.125/

ACCOUNT OF THE ACCIDENT

The potassium iodide supplies in a Harrisburg warehouse.

ACCOUNT OF THE ACCIDENT

TMI personnel cleaning up the contaminated auxiliary building.

140

ACCOUNT OF THE ACCIDENT

The initial cleanup began in April. Using a system called EPICOR-I, Met Ed began decontaminating pre-accident water stored in the TMI-1 auxiliary building, which contains low levels of radioactivity (less than one microcurie per milliliter). Efforts to decontaminate the TMI-2 diesel generation building began in April and work on the auxiliary and fuel handling buildings got under way in May. This involves mostly dry and wet vacuuming, mopping, and wiping of radioactive areas to remove the contamination -- a task that requires special clothing and respirators to protect workers.

The accident and its subsequent cleanup already have produced a variety of solid, slightly radioactive wastes, such as clothing, rags, ion-exchange resins, swipes, and contaminated air filters. To date, 12 truckloads of these wastes have been hauled to Richland, Washington, and buried at a commercial disposal site.

But the more difficult aspects of decontamination -- both technically and politically -- lie ahead. Met Ed has asked the NRC for permission to release the krypton-85 in the air of the containment building into the atmosphere in controlled bursts. The releases would come over a 2-month period to ensure that off-site radiation does not exceed the NRC's limits for routine operation of a nuclear power plant.

Much of the contaminated water left from the accident -- some 600,000 gallons pooled in the containment building and about 90,000 gallons in the reactor coolant system -- contains high levels of radioactivity (in excess of 100 microcuries per milliliter). Met Ed had stored 380,000 gallons of water containing intermediate levels of radioactivity (1 to 100 microcuries per milliliter) in several TMI-2 auxiliary building tanks. Over the summer, the utility installed a system called EPICOR-II to treat this water. NRC approved its use, provided that the resins used to remove radioactive materials from the water were solidified before shipment from the Island to a disposal site. Met Ed began decontaminating the intermediate water in mid-October.

Until radioactive gases are removed from the containment building, no human entry into the sealed structure can be made. Meanwhile, detailed plans for entry and assessing conditions inside the building are being developed. Because no one knows the exact condition of the reactor vessel or its core, no detailed plans have been made for handling and removing its damaged core.

Thus, the accident at Three Mile Island, in a very real sense, continues and will continue until the years-long cleanup of TMI-2 is completed. Workers will receive additional radiation doses until the decontamination process is completed; five workers in late August, for example, received doses in excess of the NRC's quarterly limits for exposure to the skin or the extremities. And there still remains some risk to the general public that released radiation could escape from the Island.

ACCOUNT OF THE ACCIDENT

FOOTNOTES

1. For times cited in narrative, please refer to the "Catalog of Events" compiled by the Commission's Technical Assessment Task Force. This unpublished document is available with the Commission's official records in the National Archives.

2. Commission Hearing, May 19, 1979, pp. 24, 27-28. For a discussion of the mental health effects of the TMI accident, see "Report of the Behavioral Effects Task Group."

3. General Public Utilities Corporation, 1978 Annual Report. For a full discussion of the Commission's investigation of GPU and its subsidiary, Metropolitan Edison, see "Report of the Office of Chief Counsel on the Role of the Managing Utility and Its Suppliers."

4. See history section of "Report of the Office of Chief Counsel on the Nuclear Regulatory Commission."

5. See "How a Nuclear Reactor Works," a lecture presented by Commissioner Theodore B. Taylor to members of the Commission, April 26, 1979.

6. Final Safety Analysis Report, TMI-2, Vol. 4, pp. 4.2-37, 4.3-20. See also Taylor, *supra*.

7. Final Safety Analysis Report, TMI-2, Vol. 4, pp. 4.2-4, 4.2-5, 4.3-19; figures 4.2-4, 4.3-25, 4.2-26; and Vol. 7, p. 7.8-3; figure 7.8-4.

8. See Taylor, *supra*. See also technical staff analysis report on "Alternative Event Sequences."

9. Final Safety Analysis Report, TMI-2, Vol. 3, p. 3.8-38.

10. See technical staff analysis reports on "Core Damage," "Chemistry," "Thermal Hydraulics," and "Alternative Event Sequences."

11. For a complete discussion of the pilot-operated relief valve, see technical staff analysis report on "Pilot-Operated Relief Valve

ACCOUNT OF THE ACCIDENT

Design and Performance," and section on PORV failure history in "Report of the Office of Chief Counsel on the Role of the Managing Utility and Its Suppliers." See also discussion of "safety-related" items in "Report of the Office of Chief Counsel on the Nuclear Regulatory Commission."

12. Commission Hearing, May 30, 1979, pp. 10, 115, 152-153.

13. For a full discussion of Metropolitan Edison's and Babcock & Wilcox's treatment of problems associated with pilot-operated relief valves, see "Report of the Office of Chief Counsel on the Role of the Managing Utility and Its Suppliers."

14. For a discussion on operator training, see technical staff analysis report on "Selection, Training, Qualification, and Licensing of Three Mile Island Reactor Operating Personnel." See also technical staff analysis report on "Technical Assessment of Operating, Abnormal, and Emergency Procedures;" and sections on attention to experience, TMI-2 site management, and procedures in "Report of the Office of Chief Counsel on the Role of the Managing Utility and Its Suppliers;" and section on operator training and licensing in "Report of the Office of Chief Counsel on the Nuclear Regulatory Commission."

15. Id.

16. For a full discussion of the control room, see technical staff analysis report on "Control Room Design and Performance." See also "Report of the Office of Chief Counsel on the Role of the Managing Utility and Its Suppliers" for a discussion of the TMI-2 control room design history and design performance during the accident; and "Report of the Office of Chief Counsel on the Nuclear Regulatory Commission" on the NRC's consideration of human factors in design review during plant licensing.

17. Commission Hearing, May 30, 1979, p. 168.

18. See technical staff analysis report on "Condensate Polishing System." See also section on attention to experience in "Report of the Office of Chief Counsel on the Role of the Managing Utility and Its Suppliers."

19. For a discussion on operator training, see technical staff analysis report on "Selection, Training, Qualification, and Licensing of Three Mile Island Reactor Operating Personnel." See also technical staff analysis report on "Technical Assessment of Operating, Abnormal, and Emergency Procedures;" and sections on attention to experience, TMI-2 site management, and procedures of "Report of the Office of Chief Counsel on the Role of the Managing Utility and Its Suppliers;" and section on operator training and licensing in "Report of the Office of Chief Counsel on the Nuclear Regulatory Commission."

20. Commission Hearing, May 30, 1979, p. 168.

ACCOUNT OF THE ACCIDENT

21. See technical staff analysis report on "Technical Assessment of Operating, Abnormal, and Emergency Procedures." See also discussion of pressurizer level and "going solid" in procedures section of "Report of the Office of Chief Counsel on the Role of the Managing Utility and Its Suppliers."

22. Commission Hearing, May 30, 1979, p. 194.

23. Commission Hearing, May 30, 1979, pp. 114-115; Commission Hearing, May 31, 1979, p. 40.

24. See technical staff analysis report on "Closed Emergency Feedwater Valves."

25. See technical staff analysis report on "Technical Assessment of Operating, Abnormal, and Emergency Procedures." See also discussion of emergency procedures for loss of reactor coolant and for identifying the open pilot-operated relief valve in procedures section of "Report of the Office of Chief Counsel on the Role of the Managing Utility and Its Suppliers."

26. Commission Hearing, May 30, 1979, p. 128.

27. Commission Hearing, May 30, 1979, p. 129.

28. See technical staff analysis report on "Technical Assessment of Operating, Abnormal, and Emergency Procedures." See also discussion of emergency procedures for loss of reactor coolant and for identifying the open pilot-operated relief valve in procedures section of "Report of the Office of Chief Counsel on the Role of the Managing Utility and Its Suppliers."

29. Commission Hearing, May 30, 1979, p. 146.

30. See unpublished "Catalog of Events."

31. Commission Hearing, May 31, 1979, p. 38.

32. Commission Hearing, May 30, 1979, p. 187.

33. Commission Hearing, May 30, 1979, p. 161.

34. See technical staff analysis reports on "Core Damage," "Thermal Hydraulics," and "Chemistry." See also section on Met Ed's understanding of core condition on March 28 in "Report of the Office of Chief Counsel on the Role of the Managing Utility and Its Suppliers."

35. Rogers deposition, pp. 84-85.

36. Commission Hearing, May 30, 1979, pp. 119-120.

37. See technical staff analysis reports on "Core Damage" and "Chemistry."

ACCOUNT OF THE ACCIDENT

38. Final Safety Analysis Report, TMI-2, Vol. 13, p. 13A-3.

39. Commission Hearing, May 31, 1979, pp. 6-7.

40. Prepared testimony of Gary Miller, Commission Hearing, May 31, 1979: "TMI Station March 28, 1979, Incident Statement by G. P. Miller, Station Manager," pp. 3-4.

41. NRC Region I answering service log. For a complete discussion of the emergency response during the TMI accident, see "Report of the Office of Chief Counsel on Emergency Response," which is in the form of a chronology. See also "Technical Report on Emergency Preparedness and Response." For a discussion of Met Ed's management of the accident, see section on management approach to the emergency of "Report of the Office of Chief Counsel on the Role of the Managing Utility and Its Suppliers."

42. Dornsife deposition, p. 20.

43. Final Safety Analysis Report, TMI-2, Vol. 13, p. 13A-4.

44. For a discussion of containment and isolation, see technical staff analysis report on "Containment."

45. Mattson deposition, pp. 12-14. For a complete discussion of "grandfathering" and other matters related to plant licensing, see section on plant licensing in "Report of the Office of Chief Counsel on the Nuclear Regulatory Commission."

46. For a complete discussion on selection of containment isolation criteria for TMI-2, see section on containment isolation criteria in "Report of the Office of Chief Counsel on the Role of the Managing Utility and Its Suppliers."

47. See technical staff analysis report on "Containment."

48. Commission Hearing, May 31, 1979, p. 51.

49. Commission staff notes of Pintek interview, pp. 1-2.

50. For a discussion of news media coverage of the TMI accident, see "Report of the Public's Right to Information Task Force."

51. Commission Hearing, May 19, 1979, p. 125.

52. Commission Hearing, May 31, 1979, pp. 236-237.

53. In addition to unpublished "Catalog of Events," please refer to "Report of the Office of Chief Counsel on Emergency Response" for times and dates cited.

ACCOUNT OF THE ACCIDENT

54. Commission Hearing, May 19, 1979, p. 15.

55. Commission Hearing, August 2, 1979, pp. 180-183.

56. Transcript of Scranton news conference, March 28, 1979, 10:55 a.m.

57. Id. For a discussion on radiation exposure to the public during the accident and its potential health effects, see "Report of the Health Physics and Dosimetry Task Group" and "Report of the Radiation Health Effects Task Group."

58. Commission Hearing, August 2, 1979, p. 184. See also discussion of information sources in "Report of the Public's Right to Information Task Force."

59. See technical staff analysis reports on "Thermal Hydraulics," "Chemistry," and "Technical Assessment of Operating, Abnormal, and Emergency Procedures." See also discussion of rapid depressurization in emergency response section of "Report of the Office of Chief Counsel on the Nuclear Regulatory Commission."

60. For a discussion of the radiation exposure to workers during the accident, see "Report of the Health Physics and Dosimetry Task Group." For a discussion on NRC requirements for Met Ed's provisions for worker protection, see "Report of the Public Health and Epidemiology Task Group."

61. Commission Hearing, May 31, 1979, p. 57.

62. Commission Hearing, May 31, 1979, p. 58.

63. Rogers deposition, p. 114.

64. Commission Hearing, May 31, 1979, p. 59.

65. For a discussion on radiation exposure to the public during the accident and its potential health effects, see "Report of the Health Physics and Dosimetry Task Group" and "Report of the Radiation Health Effects Task Group."

66. Governor's log, March 28, 1979, p. 3.

67. Transcript of Scranton news conference, March 28, 1979, 4:30 p.m.

68. Id.

69. Commission Hearing, May 19, 1979, pp. 37, 41.

70. Dauphin County Office of Emergency Preparedness log, March 29, 1979.

71. For a discussion of radiation monitoring during the accident, see "Report of the Health Physics and Dosimetry Task Group."

ACCOUNT OF THE ACCIDENT

72. MacLeod deposition, pp. 21-26.

73. Robbins deposition, pp. 36-37.

74. MacLeod deposition, pp. 29-36.

75. Reilly interview, pp. 92-94; Gallina deposition, pp. 56-57; Smith deposition, p. 36; and Gerusky interview, pp. 24-25. See also "Report of the Public's Right to Information Task Force."

76. NRC telephone transcript, March 29, 1979, 02-228-CH6/KD-2-6.

77. Pennsylvania DER news release, March 29, 1979. See also "Report of the Public's Right to Information Task Force" and "Report of the Office of Chief Counsel on Emergency Response."

78. Higgins deposition, pp. 42, 45-46; Gallina deposition, p. 60; Critchlow interview, first and second tape, pp. 15-18.

79. Barrett deposition, pp. 39-50.

80. Commission Hearing, August 2, 1979, pp. 316.

81. Commission Hearing, May 31, 1979, pp. 172-181.

82. Commission Hearing, August 2, 1979, pp. 298-299.

83. Id., pp. 303-316.

84. Collins deposition, pp. 70-72; Scranton interview, p. 54; Gerusky deposition, pp. 53-54; Dornsife deposition, p. 76; NRC telephone transcript, March 30, 1979, 03-019-CH2.20 SW-10.

85. Molloy deposition, pp. 48-49; Commission Hearing, August 2, 1979, pp. 10-11.

86. Commission Hearing, May 31, 1979, pp. 257-258.

87. Thornburgh deposition, p. 77.

88. Commission Hearing, August 21, 1979, p. 10.

89. Gerusky deposition, pp. 64-65, 67.

90. Commission Hearing, August 21, 1979, p. 25.

91. Transcript of Herbein news conference, March 30, 1979, 11:00 a.m.

92. See "Report of the Public's Right to Information Task Force."

ACCOUNT OF THE ACCIDENT

93. Watson deposition, pp. 73-76, 90-91.

94. Commission Hearing, May 19, 1979, p. 221.

95. Commission Hearing, May 19, 1979, p. 38. For a discussion of the mental health effects of the TMI accident, see "Report of the Behavioral Effects Task Group."

96. Commission Hearing, May 19, 1979, p. 127.

97. Henderson deposition, pp. 71-73; Commission Hearing, August 2, 1979, pp. 12-18. For full discussion on emergency planning and response, see "Technical Report on Emergency Preparedness and Response," "Report of the Public Health and Epidemiology Task Group," "Report of the Office of Chief Counsel on Emergency Preparedness," and "Report of the Office of Chief Counsel on Emergency Response." For dates and times in the following section, refer to the latter report.

98. Dieckamp deposition, pp. 129-136.

99. Denton deposition, pp. 101-102. For a complete discussion of the events concerning the hydrogen bubble, see "Report of the Office of Chief Counsel on Emergency Response." See also technical analysis of hydrogen production in technical staff analysis report on "Chemistry."

100. Knouse interview, pp. 71-73; transcript of Denton-Thornburgh news conference, March 30, 1979.

101. Waldman interview, pp. 68-69.

102. Watson deposition, pp. 52-57; Eidenberg deposition, p. 47-48.

103. Commission Hearing, August 2, 1979, pp. 194-195.

104. For a complete recount of the potassium iodide story, see "Report of the Public Health and Epidemiology Task Group" and "Report of the Office of Chief Counsel on Emergency Response."

105. Villforth deposition, pp. 33-35; "Chronology of Events at HEW Regarding TMI, 3/28/79 through 4/30/79."

106. For a complete discussion of the events concerning the hydrogen bubble, see "Report of the Office of Chief Counsel on Emergency Response." See also technical analysis of hydrogen production in technical staff analysis report on "Chemistry."

107. Mattson deposition, p. 184.

108. Id., p. 186.

109. Commission Hearing, August 22, 1979, pp. 294-296. See also Mattson deposition, pp. 178-180.

ACCOUNT OF THE ACCIDENT

110. Mattson deposition, p. 179.

111. Id., pp. 190-191.

112. For a full discussion on this incident, see "Report of the Public's Right to Information Task Force."

113. Transcript of Denton-Thornburgh news conference, March 31, 1979, Part 1, p. 3.

114. For a more detailed discussion of HEW's activities during the accident, see "Report of the Office of Chief Counsel on Emergency Response" and "Report of the Public Health and Epidemiology Task Group."

115. Molloy deposition, pp. 81-100. See also Commission Hearing, August 2, 1979, p. 16.

116. Mattson deposition, p. 192.

117. Mattson interview, cassette 16, parts 5 and 6, pp. 34-35.

118. Id., cassette 17, parts 7 and 8, p. 4.

119. For a discussion on the NRC commissioners and their role in the management of the agency and during the emergency, see "Report of the Office of Chief Counsel on the Nuclear Regulatory Commission."

120. Transcript of Denton news conference, April 2, 1979. For a discussion of NRC public statements on the hydrogen bubble problems, see "Report of the Public's Right to Information Task Force."

121. Mattson interview, cassette 17, parts 7 and 8, p. 7. See also "Report of the Public's Right to Information Task Force."

122. For a complete recount of the potassium iodide story, see "Report of the Public Health and Epidemiology Task Group" and "Report of the Office of Chief Counsel on Emergency Response."

123. For discussion of Met Ed's recovery efforts, see section on TMI-2 recovery program in "Report of the Office of Chief Counsel on the Role of the Managing Utility and Its Suppliers" and technical staff analysis report on "Recovery."

124. Molloy deposition, pp. 115-117. See also "Report of the Office of Chief Counsel on Emergency Response" and "Technical Staff Report on Emergency Preparedness and Response."

125. For a discussion of TMI-2's recovery program, see technical staff analysis report on "Recovery" and the TMI-2 recovery program section of "Report of the Office of Chief Counsel on the Role of the Managing Utility and Its Suppliers."

APPENDICES
EXECUTIVE ORDER 12130

PRESIDENT'S COMMISSION ON THE ACCIDENT AT
THREE MILE ISLAND

By the authority vested in me as President by the Constitution of the United States of America, and in order to provide, in accordance with the provisions of the Federal Advisory Committee Act (5 U.S.C. App. 1), an independent forum to investigate and explain the recent accident at the nuclear power facility at Three Mile Island in Pennsylvania, it is hearby ordered as follows:

1-1. Establishment.

1-101. There is established the President's Commission on the Accident at Three Mile Island.

1-102. The membership of the Commission shall be composed of not more than twelve persons appointed by the President from among citizens who are not full time officers or employees in the Executive Branch. The President shall designate a Chairman from among the members of the Commission.

1-2. Functions.

1-201. The Commission shall conduct a comprehensive study and investigation of the recent accident involving the nuclear power facility on Three Mile Island in Pennsylvania. The study and investigation shall include:

(a) a technical assessment of the events and their causes;

(b) an analysis of the role of the managing utility;

(c) an assessment of the emergency preparedness and response of the Nuclear Regulatory Commission and other Federal, state and local authorities;

EXECUTIVE ORDER

(d) an evaluation of the Nuclear Regulatory Commission's licensing, inspection, operation and enforcement procedures as applied to this facility;

(e) an assessment of how the public's right to information concerning the events at Three Mile Island was served and of the steps which should be taken during similar emergencies to provide the public with accurate, comprehensible and timely information; and

(f) appropriate recommendations based upon the Commission's findings.

1-202. The Commission shall prepare and transmit to the President and to the Secretaries of Energy and Health, Education and Welfare a final report of its findings and recommendations.

1-3. <u>Administration</u>.

1-301. The Chairman of the Commission is authorized to appoint and fix the compensation of a staff of such persons as may be necessary to discharge the Commission's responsibilities, subject to the applicable provisions of the Federal Advisory Committee Act and Title 5 of the United States Code.

1-302. To the extent authorized by law and requested by the Chairman of the Commission, the General Services Administration shall provide the Commission with necessary administrative services, facilities, and support on a reimbursable basis.

1-303. The Department of Energy and the Department of Health, Education, and Welfare shall, to the extent permitted by law and subject to the availability of funds, provide the Commission with such facilities, support, funds and services, including staff, as may be necessary for the effective performances of the Commission's functions.

1-304. The Commission may request any Executive agency to furnish such information, advice or assistance as it deems necessary to carry out its functions. Each such agency is directed, to the extent permitted by law, to furnish such information, advice or assistance upon request by the Chairman of the Commission.

1-305. Each member of the Commission may receive compensation at the maximum rate now or hereafter prescribed by law for each day such member is engaged in the work of the Commission. Each member may also receive travel expenses, including per diem in lieu of subsistence (5 U.S.C. 5702 and 5703).

1-306. The functions of the President under the Federal Advisory Committee Act which are applicable to the Commission, except that of reporting annually to the Congress, shall be performed by the Administrator of General Services.

EXECUTIVE ORDER

1-4. Final Report and Termination.

    1-401. The final report required by Section 1-202 of this Order shall be transmitted not later than six months from the date of the Commission's first meeting.

    1-402. The Commission shall terminate two months after the transmittal of its final report.

                                                /s/ Jimmy Carter

THE WHITE HOUSE
April 11, 1979

COMMISSION OPERATIONS & METHODOLOGY

This appendix will appear in the permanent edition of the report.

COMMISSIONERS' BIOGRAPHIES

Babbitt, Bruce. Governor of Arizona, 1978- . Born: June 27, 1938, Los Angeles, Calif. Education: Notre Dame University (BA, 1960); University of Newcastle, England (MS, 1963); Harvard Law School (LLB, 1965). Experience: special assistant to director, VISTA, 1966-67; attorney, Brown & Bain, Phoenix, Ariz., 1967-74; attorney general of Arizona, 1974-78. Honors & Awards: Marshall Scholar, 1960-62; Thomas Jefferson Award, 1979, Society of Professional Journalists-Sigma Delta Chi. Publications: Grand Canyon: An Anthology (1978); Color and Light: The Southwest Canvases of Louis Akin (1974). Memberships: National Governors' Association (chairman, Subcommittee on Public Protection); Four Corners Regional Commission; former member, National Association of Attorneys General. Activities: Advisory Commission on Intergovernmental Relations; chairman, Southwest Border Regional Commission; Advisory Committee, Kennedy School of Government, Harvard University.

Haggerty, Patrick Eugene. General Director and Honorary Chairman, Texas Instruments, Inc., Dallas, Tex., 1976- . Born: March 17, 1914, Harvey, N.D. Education: Marquette University (BS, 1936). Experience: Badger Carton Co., Milwaukee, Wis., 1935-42: production manager, 1935-39, assistant general manager, 1939-42; lieutenant, U.S. Naval Reserve at Bureau of Aeronautics, U.S. Dept. of the Navy, 1942-45; Texas Instruments, Inc., 1945-76: general manager, Laboratory and Manufacturing Division, 1945-51, executive vice president and director, 1951-58, president, 1958-66, chairman, 1966-76, retired, 1976. Honors & Awards: Medal of Honor, 1967, Electronic Industries Association; Founders Award, 1968, Institute of Electrical and Electronics Engineers; Industrial Research Institute Medalist, 1969; John Fritz Medalist, 1971; Alumnus of the Year, 1972, Marquette University; Wema Medal of Achievement, 1972; Henry Laurence Gantt Medal, 1975; honorary doctorates from: St. Mary's University, 1959, Marquette University, 1960, Polytechnic Institute of Brooklyn, 1962, University of Dallas, 1964, North Dakota State University, 1967, Catholic University, 1971, Rensselaer Polytechnic Institute, 1972, University of Notre Dame, 1974. Publications: Management Philosophies and Practices of Texas Instruments (1965), The Productive Society (1973). Memberships: Rockefeller University (chairman, Board of Trustees); University of Dallas (Board of Trustees and Executive

COMMISSIONERS' BIOGRAPHIES

Committee); Institute of Electrical and Electronics Engineers (Fellow); National Academy of Engineering (elected, 1965); American Association for the Advancement of Science (Fellow); National Security Industrial Association (life member and former vice chairman, Board of Trustees); Texas Academy of Science (life member); The Business Council. Activities: vice chairman, Defense Science Board, 1965-67; National Commission on Technology, Automation, and Economic Progress, 1968; Presidential Science Advisory Committee, 1970-71; chairman, National Council on Educational Research, 1973-74; Executive Committee, Tri-Lateral Commission, 1973-76; Board of Governors, U.S. Postal Service, 1972-73.

Kemeny, John G. President, Dartmouth College, Hanover, N.H., 1970- . Born: May 31, 1926, Budapest, Hungary; came to U.S. 1940; naturalized U.S. citizen, 1945. Education: Princeton University (BA, 1947; PhD, 1949). Experience: assistant, Theoretical Division, Manhattan Project, U.S. Dept. of the Army, Los Alamos Scientific Laboratory, N. Mex., 1945-46; research assistant to Albert Einstein, Institute for Advanced Study, 1948-49; Princeton University, 1949-53: Fine Instructor in Mathematics, 1949-51, assistant professor of philosophy, 1951-53; Dartmouth College, 1953-70: professor of mathematics, 1953-70, chairman of Mathematics Department, 1955-67, Albert Bradley Third Century Professor, 1969-70. Honors & Awards: Priestley Award, 1976; honorary doctorates from: Middlebury College, 1965, Columbia University, 1971, Princeton University, 1971, University of New Hampshire, 1972, Boston College, 1973, University of Pennsylvania, 1975, Colby College, 1976, Bard College, 1978, Lafayette College, 1978. Publications: A Philosopher Looks at Science (1959), Man and the Computer (1972), Random Essays, numerous articles; co-author: Basic Programming (1968), Denumerable Markov Chains (1966), Finite Mathematics with Business Applications (1962), Finite Markov Chains (1960), Mathematical Models in the Social Sciences (1962), Finite Mathematical Structures (1958), Introduction to Finite Mathematics (1957); contributor, Encyclopaedia Britannica; associate editor, Journal of Mathematical Analysis and Applications, 1959-70. Memberships: Phi Beta Kappa; Association for Symbolic Logic (consulting editor, 1950-59); Mathematical Association of America (chairman, New England Section, 1959-60; Board of Governors, 1960-63; chairman, Panel on Teacher Training, 1961-63; chairman, Panel on Biological and Social Sciences, 1963-64); American Mathematical Society; American Philosophical Association; American Academy of Arts and Sciences; Sigma Xi (national lecturer, 1967); National Council of Teachers of Mathematics; American Association for the Advancement of Science. Activities: consultant, Rand Corp., 1953-70; consultant, Educational Research Council of Greater Cleveland, 1959-70; chairman, U.S. Commission on Mathematics Instruction, 1958-60; National Research Council, 1963-66; vice chairman, Advisory Committee on Computing, National Science Foundation, 1968-69; delivered Vanuxem Lectures, Princeton University, 1974; National Commission on Libraries and Information Science, 1971-73; Advisory Committee to Regional Director, U.S. Dept. of Health, Education, and Welfare, 1971-73; trustee, Foundation Center, 1970-76; trustee, Carnegie Foundation for Advancement of Teaching, 1972-78; director, Council for Financial Aid to Education, 1976-79; director, Honeywell, Inc., 1978-79.

COMMISSIONERS' BIOGRAPHIES

Lewis, Carolyn Diana. Associate Professor and Coordinator of the Broadcast Division, Columbia University Graduate School of Journalism, New York, N.Y., 1978-   . Born: Sept. 8, 1931, New York, N.Y. Education: University of Arizona (BA, 1951); University of Oslo, Norway (1951); University of Sydney, Australia (1956-58). Experience: reporter, United Press International, Sydney, 1954-56; radio news commentator, Macquarie Network, Australia, 1958-65; television news commentator, ATN, Sydney, 1958-65; reporter and columnist, Daily Mirror, Sydney, 1959-65; reporter, Washington Post, 1965-68; Capitol Hill correspondent, WTOP-TV and radio, 1968-74; moderator, "Meeting of the Minds," WNBC, 1966-68; Capitol Hill correspondent, Television News, Inc., 1974; correspondent, impeachment hearings, PBS (National Public Affairs Center for Television), 1974-75; associate professor of journalism, Boston University, 1975-78. Honors & Awards: Gavel Award, American Bar Association. Publications: regular contributor, Op-Ed, The New York Times; Washington Post; Reader's Digest; TV Guide; chapter in 1979 Dupont-Columbia University Survey of Broadcast Journalism. Memberships: Phi Kappa Phi; Delta Sigma Rho.

Marks, Paul Alan. Vice President for Health Sciences, 1973-   , Frode Jensen Professor of Medicine, 1974- ; Professor of Human Genetics and Development, 1969- , Director, Cancer Center, 1973-   , Columbia University, New York, N.Y. Born: Aug. 16, 1926, New York, N.Y. Education: Columbia College, Columbia University (AB, 1945); College of Physicians and Surgeons, Columbia University (MD, 1949). Experience: senior investigator, National Institutes of Health, 1953-55; Columbia University, 1952- : Fellow, College of Physicians and Surgeons, 1952-53, instructor in medicine, 1955-56, associate in medicine, 1956-57, assistant professor of medicine, 1957-61, associate professor of medicine, 1961-67, professor of medicine, 1967-74, chairman, Dept. of Human Genetics and Development, 1969-70, dean, Faculty of Medicine, 1970-73; visiting attending physician, Francis Delafield Hospital, 1959-75; visiting scientist, Laboratory of Cellular Biochemistry, Pasteur Institute, Paris, France, 1961-62; Presbyterian Hospital, 1962- : associate attending physician, 1962-67, attending physician, 1967-   . Honors & Awards: Phi Beta Kappa; Columbia University: Janeway Prize, 1949, Joseph Mather Smith Prize, 1959, Stevens Triennial Prize, 1960, Bicentennial Medal, 1968; Swiss-American Foundation Award in Medical Research, 1965; Fellow, American Academy of Arts and Sciences, 1972; Institute of Medicine, National Academy of Sciences, 1972. Publications: numerous articles; Editorial Board, 1964-71, associate editor, 1976-77, editor-in-chief, 1978-   , Blood, Journal of the American Society of Hematology; editor, 1967-71, associate editor, 1971-72, Journal of Clinical Investigation; consulting editor, Blood Cells, 1974-   . Memberships: American Association for the Advancement of Science; American Association for Cancer Research; American College of Physicians; American Society of Hematology (chairman, Committee on Scientific Affairs, 1970, chairman, Publications Committee, 1973-74); American Society of Human Genetics (Program Committee, 1963); Association of American Physicians; American Society for Clinical Investigation (president, 1971-72); National Academy of Sciences (elected, 1973). Activities: Advisory Panel, National Science Foundation, 1964-67; Organizing Committee, International School for Developmental Biology, NATO, 1971-72; Delos Conferences,

COMMISSIONERS' BIOGRAPHIES

Athens, Greece, 1971-73; chairman, Hematology Training Grants Committee, National Institutes of Health, 1971-73; chairman, Executive Committee, Division of Medical Sciences, National Research Council, National Academy of Sciences, 1973-76; Founding Committee, Radiation Effects Research Foundation, Japan, 1975-77; President's Panel on Biomedical Research, 1975-76; President's Cancer Panel, 1976-   ; trustee, St. Luke's Hospital Center, 1970-   ; trustee, Roosevelt Hospital, 1970-   ; trustee, Presbyterian Hospital, 1972-   ; Board of Directors, Keystone Life Sciences Center, Colo., 1976-   ; director, Charles H. Revson Foundation, Inc., 1976-   ; Board of Trustees, Metpath Institute for Medical Education, 1977-   ; director, Dreyfus Leverage Fund, 1978-   ; director, Pfizer Inc., 1978-   .

Marrett, Cora Bagley. Professor of Sociology and Afro-American Studies, University of Wisconsin, Madison, 1979-   . Born: June 15, 1942, Richmond, Va. Education: Virginia Union University (BA, 1963), University of Wisconsin (MA, 1965; PhD, 1968). Experience: assistant professor of sociology, University of North Carolina, 1968-69; Western Michigan University, 1969-74: assistant professor of sociology, 1969-72, associate professor of sociology, 1972-74; associate professor, Departments of Sociology and Afro-American Studies, University of Wisconsin, 1974-79. Honors & Awards: Summer Stipend for Younger Humanists, National Endowment for the Humanities, 1972; Resident Fellow, National Academy of Sciences, 1973-74; Fellowship, Center for Advanced Study in the Behavioral Sciences, 1976-77. Publications: numerous articles in technical journals. Memberships: American Sociological Association (chairwoman, Committee on the Status of Women in Sociology, 1972-74; Editorial Board, American Sociologist, 1972-75; Minority Fellowship Committee, 1973-74); Southwestern Social Science Association (Editorial Board, Social Science Quarterly, 1972-   ); Social Science Research Council (Board of Directors, 1973-80; Executive Committee, 1975-80). Activities: U.S. Army Scientific Advisory Panel: consultant, 1976-77, Research Advisory Group, Institute for the Behavioral and Social Sciences, 1976-77; American Association for the Advancement of Science: conference consultant, "Minority Women in Science and Engineering," 1975, "Women in Basic Research Careers," 1977, conference co-organizer, "Minorities in Science," 1976; National Academy of Sciences, National Research Council: Social Science Panel, Advisory Committee to the U.S. Dept. of Housing and Urban Development, 1970-71, chairwoman, workshop, "Issues in the Employment of Women," 1974; National Science Foundation: Advisory Panel, Minority Institutions Science Improvement Program, 1977, Advisory Committee, Program on Ethics and Values in Science and Technology, 1977-78.

McBride, Lloyd. International President, United Steelworkers of America, Pittsburgh, Pa., 1977-   ; Vice President, AFL-CIO, 1977-   . Born: March 9, 1916, Farmington, Mo. Experience: member, Steel Workers Organizing Committee, 1936; organizer and Negotiating Committee, Local 1295, St. Louis, Mo., 1936; president, Local 1295, 1938-40; president, St. Louis Industrial Union Council of CIO, 1940-42; president, Missouri State CIO Industrial Union Council, 1942-44; U.S. Naval Reserve, 1944-46; United Steelworkers of America, AFL-CIO, 1946-   : staff union representative, 1946-58, subdistrict director, 1958-65, director, District 34 (Mo., Kan., Neb., Iowa, southern

COMMISSIONERS' BIOGRAPHIES

Ill.), (secretary and chairman, basic steel negotiation, Armco Steel Corp.; chairman, multi-plant bargaining, American Steel Foundries), 1965-77. Honors & Awards: Honorary Fellow, Truman Library Institute. Memberships: National Urban Coalition (Board of Directors); American Arbitration Association (director); Americans for Energy Independence (Board); National Center for Resource Recovery (Board of Directors); National Society for the Prevention of Blindness (Board of Directors); Salvation Army (advisory member). Activities: chairman, Foundry and Forgings Industry Conference; chairman, Lead Workers Conference; delegate, International Metalworkers Federation, Geneva, Switzerland; Labor Policy Advisory Committee, U.S. Dept. of Labor; vice president, American Immigration and Citizenship Conference; Advisory Committee for Trade Negotiations; National Commission on Air Quality; President's Committee on Employment of the Handicapped.

McPherson, Harry C., Jr. Partner; Verner, Liipfert, Bernhard, & McPherson; Washington, D.C.; 1969- . Born: Aug. 22, 1929, Tyler, Tex. Education: University of the South (BA, 1949), Columbia University (1949-50), University of Texas Law School (LLB, 1956). Experience: U.S. Senate Democratic Policy Committee, 1956-63: assistant general counsel, 1956-59, associate general counsel, 1959-61, general counsel, 1961-63; deputy under secretary of the Army for international affairs and special assistant to the secretary for civil functions, 1963-64; assistant secretary of state for educational and cultural affairs, 1964-65; special assistant and counsel (1965-66) and special counsel (1966-69) to President Lyndon B. Johnson. Honors & Awards: Distinguished Civilian Service Award, 1964, U.S. Dept. of the Army; Honorary DCL, 1965, University of the South; Arthur S. Fleming Award, 1968. Publications: A Political Education (1972); Editorial Board, Foreign Affairs; Publications Committee, The Public Interest. Memberships: N.Y. Council on Foreign Relations (Board of Directors, 1974-77); Maryland Inquiry Panel. Activities: John F. Kennedy Center for Performing Arts: vice chairman, 1969-76, general counsel, 1977- ; Board of Trustees, Woodrow Wilson International Center for Scholars, 1969-74; chairman, Task Force on Domestic Policy, Democratic Advisory Council of Elected Officials, 1974-76.

Peterson, Russell W. President and Chief Executive Officer, National Audubon Society, New York, N.Y., 1979- . Born: Oct. 3, 1916, Portage, Wis. Education: University of Wisconsin (BS, 1938; PhD, 1942). Experience: E.I. DuPont de Nemours & Co., Inc., 1942-69: research director, Textile Fabrics Division, 1954-55 and 1956-59, merchandising manager, Textile Fibers, 1955-56, director, New Products Division, Textile Fibers, 1959-62, director, Research and Development Division, 1963-69; governor, State of Delaware, 1969-73; chairman, President's Council on Environmental Quality, 1973-76; founding president, New Directions, Washington, D.C., 1976-77; director, Office of Technology Assessment, U.S. Congress, 1978-79. Honors & Awards: Conservationist-of-the-Year, 1971, National Wildlife Federation; Gold Medal Award, 1971, World Wildlife Fund; Annual Award, 1971, Commercial Development Association; Parson's Award, 1974, American Chemical Society; Annual Award, 1977, National Audubon Society; Proctor Prize of Sigma Xi, 1978; Distinguished Citizen Award, National Municipal League; Honorary Fellow, Textile Research Institute;

COMMISSIONERS' BIOGRAPHIES

honorary doctorates from: Williams College, 1975, Stevens Institute of Technology, 1979, Butler University, 1979, Springfield College, 1979. Publications: numerous articles in Chemical and Engineering News, American Scientist, Smithsonian, Industry Week, Harvard Business Review, Bioscience, The New York Times, Washington Post, Congressional Record. Memberships: Phi Beta Kappa; Sigma Xi; North-South Roundtable; International Institute for Environment and Development; American Chemical Society; Federation of American Scientists; American Ornithological Union; American Institute of Chemists (Fellow); American Association for the Advancement of Science (director); Alliance to Save Energy (director); The World Wildlife Fund (director). Activities: regional vice president, National Municipal League, 1968- ; Board of Directors, Textile Research Institute, 1956-69 (chairman, Executive Committee, 1959-61; chairman, Board of Directors, 1961-63); co-chairman, Save Our Seas; director, Tri-County Conservancy of the Brandywine; chairman, National Education Commission of the States, 1970; chairman, National Advisory Commission on Criminal Justice Standards and Goals, 1971-73; chairman, Southern States Nuclear Board; chairman, Delaware River Basin Commission, 1971-72; chairman, Executive Committee, National Commission on Critical Choices for Americans, 1973-74; vice chairman, U.S. Delegation to the U.N. World Population Conference, 1974; vice chairman, U.S. Delegation to the U.N. World Conference on Human Settlements, 1976.

Pigford, Thomas H. Professor of Nuclear Engineering and Chairman, Department of Nuclear Engineering, University of California, Berkeley, 1959- . Born: April 21, 1922, Meridian, Miss. Education: Georgia Institute of Technology (BS, 1943); Massachusetts Institute of Technology (SM, 1948; ScD, 1952). Experience: Massachusetts Institute of Technology, 1946-57: instructor of chemical engineering, 1946-47, assistant professor of chemical and nuclear engineering, 1950-55, associate professor, 1955-57; director, Graduate School of Engineering Practice, Oak Ridge, Tenn., 1950-52; senior development engineer, Aqueous Homogenous Reactor Project, Oak Ridge National Laboratory, 1952; director of engineering, director of nuclear reactor projects, assistant director of research laboratory, General Atomic Division, General Dynamics Corp., San Diego, Calif., 1957-59. Honors & Awards: E. I. DuPont de Nemours Fellowship, 1948-50; Outstanding Young Man of Greater Boston, 1956; Arthur H. Compton Award, 1971, American Nuclear Society; Fellowship, Japan Society for the Promotion of Science, 1975; Founders Award, 1978, American Nuclear Society. Publications: co-author, Nuclear Chemical Engineering (1957); numerous research papers and articles. Memberships: National Academy of Engineering (elected, 1976); American Nuclear Society (charter member, Fellow, former director); National Atomic Safety and Licensing Board Panel (1963-72); American Association for the Advancement of Science; American Institute of Chemical Engineering; American Institute of Mechanical Engineers; Committee on Radioactive Waste Management, National Research Council, National Academy of Sciences. Activities: teaching, research, consultant to government and various industries.

Taylor, Theodore B. Visiting Lecturer with rank of Professor, Mechanical and Aerospace Engineering Department, Princeton University, 1976- . Born: July 11, 1925, Mexico City, Mexico (parents American

COMMISSIONERS' BIOGRAPHIES

citizens). Education: California Institute of Technology (BS, 1945); Cornell University (PhD, 1954). Experience: U.S. Naval Reserve, 1942-46; theoretical physicist, University of California Radiation Laboratory, Berkeley, 1946-49; staff member, Los Alamos Scientific Laboratory, 1949-56; technical director of Project Orion, senior research advisor, General Atomic Division, General Dynamics Corp., San Diego, Calif., 1956-64; deputy director (scientific), Defense Atomic Support Agency, U.S. Dept. of Defense, 1964-66; chairman of the board and founder, International Research and Technology Corp., 1967-76. Honors & Awards: Ernest O. Lawrence Memorial Award, 1965. Publications: co-author: The Restoration of the Earth (1973), Nuclear Theft: Risks and Safeguards (1974), Nuclear Proliferation (1977), Energy: The Next Twenty Years (1979), and author of numerous articles in technical journals and popular media. Memberships: American Association for the Advancement of Science, American Physical Society, International Solar Energy Society. Activities: consultant: Air Force Science Advisory Board, 1955-58, Los Alamos Scientific Laboratory, 1956-64, Aerospace Corp., 1960-61, Atomic Energy Commission, 1966-70, Defense Atomic Support Agency, 1966-69, Rockefeller Foundation, 1977-79; chairman, Los Alamos Study Group, Air Force Space Study Commission, 1961; Voluntary Speaker Program in Asia, U.S. Information Service, 1977-79.

Trunk, Anne D. Resident, Middletown, Pa.; married, six children. Born: Dec. 31, 1934, New York, N.Y. Activities: Middletown Civic Club (president, 1971-72); Middletown Women's Club; Advisory Board, L. J. Fink Elementary School, Middletown, Pa.; Welcome Committee, Terre Haute, Ind.

STAFF LIST

OFFICE OF THE CHIEF COUNSEL

Stanley M. Gorinson, Chief Counsel

Emergency Preparedness Team
    Charles A. Harvey, Jr., Team Leader
    Ruth Dicker
    Eric Pearson

Role of the Managing Utility and Its Suppliers Team
    Winthrop A. Rockwell, Team Leader
    Joan L. Goldfrank
    Michael R. Hollis

Role of the Nuclear Regulatory Commission Team
    Kevin P. Kane, Deputy Chief Counsel and Team Leader
    Stan M. Helfman, Team Co-leader
    Gary M. Sidell

Law Clerks
    Jeffrey Klein
    Kenneth D. Noel

Legal Assistants
    Thomas S. Blood, Jr.
    Louis F. Cooper
    Mary Ann Hanlon
    Sondra M. Korman
    Leslie A. Moushey
    Susan R. Paisner
    Karen Randall
    Dan W. Reicher
    Daniel J. Sherman
    Claudia A. Velletri

STAFF LIST

OFFICE OF THE DIRECTOR OF TECHNICAL STAFF

Vincent L. Johnson, Director of Technical Staff

Technical Assessment Task Force
- Leonard Jaffe, Head
- William M. Bland, Jr.
- Robert English
- Ronald M. Eytchison
- Dwight Reilly
- William Stratton
- Jasper L. Tew

Public Health and Safety Task Force
- Jacob I. Fabrikant, Head
    - Lloyd Corwin, Advisory/Administrative Associate
- John A. Auxier, Head, Health Physics and Dosimetry Task Group
- George W. Casarett, Head, Radiation Health Effects Task Group
- Paul M. Densen, Head, Public Health and Epidemiology Task Group
- Bruce P. Dohrenwend, Head, Behavioral Effects Task Group

Emergency Preparedness and Response Task Force
- Russell R. Dynes, Head
- Quinten Johnson
- Arthur H. Purcell
- Robert Stallings
- Philip Stern
- Dennis E. Wenger

Public's Right to Information Task Force
- David M. Rubin, Head
- Holly A. Chaapel
- Ann Marie Cunningham
- Nadyne G. Edison
- Mary Beth Franklin
- Sharon M. Friedman
- Wilma I. Hill
- Nancy C. Joyce
- Mary Paden
- Roy S. Popkin
- Peter M. Sandman
- Mark C. Stephens
- Mitchell Stephens
- Patricia E. Weil
- Emily Wells

STAFF LIST

    Administrative Staff
        Francis T. Hoban, Director of Administration
        Fred Bowen, Former Director of Administration
        Jeanne L. Holmberg, Administrative Officer
        Betty A. Turner, Administrative Officer
        Richard Sprince, Assistant to the Director
        Nancy D. Watson, Staff Assistant

        Barbara R. Chesnik, Research Associate/Supervisor-
         Document Control
        Madeline Pinelli, Former Supervisor-Document Control

STAFF LIST

OFFICE OF THE PUBLIC INFORMATION DIRECTOR

    Barbara Jorgenson, Public Information Director

        Cathryn M. Dickert, Congressional Relations/
       Executive Assistant to the Director

        Public Information Staff
            Wilma I. Hill, Media Associate
            Holly A. Chaapel, Public Information Assistant
            Mary Beth Franklin, Public Information Assistant

        Administrative Staff
            Wayne Fortunato-Schwandt, Administrative
             Associate/Manuscript Production
            Elsie E. Whited, Staff Assistant

        Editorial Staff
            Patrick Young, Senior Writer
            Nancy C. Joyce, Staff Writer/Media Associate
            Judy Ross Ferguson, Photo and Copy Editor
            Joseph Foote, Production Consultant

        Copy Editors
            Lisa Berger
            Nina Graybill
            Michael F. Leccese
            Vivian Noble
            Claude E. Owre
            Roger L. Wall

STAFF LIST

COMMISSION CONSULTANTS

Special Assistants to the Chairman

    Ruth LaBombard
    Elizabeth R. Dycus

Consultants to the Office of the Chief Counsel

    Dennis Bakke
    Ernest Gellhorn
    Harold Green
    Samuel W. Jensch
    Richard Stewart
    Peter Strauss

Consultants to the Technical Assessment Task Force

    Louis Baker
    Robert Burns
    Arthur M. Carr
    Paul Cohen
    Michael Corridini
    Hans Fauske
    Herb E. Feinroth
    Mario Fontana
    Frederick Forscher
    Peter Griffith
    Jerry Griffiths
    Raymond Heiskala
    George Inskeep
    John Ireland
    Walter Kirchner
    Dave Latham
    Harry Lawroski
    Saul Levy
    John Mailer
    Peter Mailer
    Peter Mast
    Margaret Mlynczak
    Frank Muller
    John Orndoff
    Matthew Opeka
    Dana Powers
    A. B. Reynolds
    Warren Rosenau
    Robert Seale
    Michael Stern
    Neil E. Todreas
    Beverly Washburn

STAFF LIST

Consultants to the Public Health and Safety Task Force

    James F. Crow
    David A. Hamburg
    Henry S. Kaplan
    Walsh McDermott

Consultants to the Health Physics and Dosimetry Task Group

    Carol D. Berger
    Charles M. Eisenhauer
    Thomas F. Gesell
    Alun R. Jones
    Mary Ellen Masterson

Consultants to the Radiation Health Effects Task Group

    Seymour Abrahamson
    William J. Bair
    Michael A Bender
    Arthur D. Bloom
    Victor P. Bond

Consultants to the Public Health and Epidemiology Task Group

    David Axelrod
    Maura S. Bluestone
    Eugene W. Fowinkle
    Kenneth G. Johnson
    Ellen W. Jones
    Raymond Seltser

Consultants to the Behavioral Effects Task Group

    Barbara S. Dohrenwend
    Stanislav V. Kasl
    George J. Warheit

Consultant to the Emergency Preparedness Task Force

    John Ruch

Consultants to the Public's Right to Information Task Force

    Alton Blakeslee
    Hillier Krieghbaum
    Howard J. Lewis

STAFF LIST

SUPPORT STAFF

Evan M. Adelson	Barrington R. Johnson
Margie A. Baker	Julie Kantrowitz
Virginia Barnes	Alicia Lucas
Estelle Barrios	Connie McDougal
Ruth Anne Beer	Dennis Melby
Elizabeth Brown	Doris C. Miller
Robin A. Brown	Jane Modatic
Wilma O. Bryan	Sharon L. Neuenberger
Sheari Carruth	Donald O'Grady
Barbara L. Cole	Ada Panciarelli
Gloria J. Coley	Kevin A. Powers
Georgia G. Davis	James P. Pryor
Simone Demers	Maggie Ruiz
Dan M. Einfalt	Sheila Elaine Saunders
Adam M. Eisgrau	Nora Schwartz
Linda Fantacci	Beth Stephens
Frederick W. Fisher	Susan Stevenson
LaJuan M. Fowler	Janet Stewart
June H. Frazier	Trisha Thompson
Ellen Glassman	Diane L. Walter
Jerrold Greenberg	Willa Weatherly
Barbara Harmon	Steven R. Weiss
Kelly L. Hilleary	Pixie J. Westhoven
Lorraine Hodge	Ronald Woerner
Martha Hollister	Mary Woolard
Ada Hungerford	Marily Woznicki

GLOSSARY

Auxiliary building - A structure housing a variety of equipment and large tanks necessary for the operation of the reactor. These include make-up pumps, the make-up and waste gas decay tanks, and the reactor coolant hold-up tanks.

Babcock & Wilcox (B&W) - The company that designed and supplied the TMI-2 reactor and nuclear steam supply system.

Background radiation - Radiation arising from natural radioactive materials always present in the environment, including solar and cosmic radiation and radioactive elements in the upper atmosphere, the ground, building materials, and the human body.

Beta particles - High-energy electrons; a form of ionizing radiation that normally is stopped by the skin, or a very thin sheet of metal.

Bureau of Radiation Protection (BRP) - A division of Pennsylvania's Department of Environmental Resources. BRP is the state's lead agency in monitoring radiation releases from nuclear plants and advises the Pennsylvania Emergency Management Agency during radiological emergencies.

Burns and Roe - Architectural and engineering firm responsible for the design of TMI-2.

Candy cane - The section of pipe carrying water from the reactor to a steam generator.

Chain reaction - A self-sustaining reaction; occurs in nuclear fission when the number of neutrons released equals or exceeds the number of neutrons absorbed plus the neutrons which escape from the reactor.

Cladding - In a nuclear reactor, the metal shell of the fuel rod in which uranium oxide pellets are stacked.

Collective dose - The sum of the individual doses received by each member of a certain group or population. It is calculated by multi-

GLOSSARY

plying the average dose per person by the number of persons within a specific geographic area. Consequently, the collective dose is expressed in person-rems. For example, a thousand people each exposed to one rem would have a collective dose of 1,000 person-rems.

<u>Condensate booster pumps</u> - Three pumps located between the condensate polisher and the main feedwater pumps.

<u>Condensate polisher</u> - A device that removes dissolved minerals from the water of the feedwater system.

<u>Condensate pumps</u> - Three pumps in the feedwater system that pump water from the condensers to the condensate polishers.

<u>Condensers</u> - Devices that cool steam to water after the steam has passed through the turbine.

<u>Containment building</u> - The structure housing the nuclear reactor; intended to contain radioactive solids, gases, and water that might be released from the reactor vessel in an accident.

<u>Control rod</u> - A rod containing material that absorbs neutrons; used to control or halt nuclear fission in a reactor.

<u>Core</u> - The central part of a nuclear reactor that contains the fuel and produces the heat.

<u>Critical</u> - Term used to describe a nuclear reactor that is sustaining a chain reaction.

<u>Curie</u> - A unit of the intensity of radioactivity in a material. A curie is equal to 37 billion disintegrations each second.

<u>Decay heat</u> - Heat produced by the decay of radioactive particles; in a nuclear reactor this heat, resulting from materials left from the fission process, must be removed after reactor shutdown to prevent the core from overheating. See <u>radioactive decay</u>.

<u>Design basis accident (DBA)</u> - Hypothetical accidents evaluated during the safety review of nuclear power reactors. Plants are required to have safeguards that will ensure that radiation releases off-site will be within NRC limits should any of these accidents occur.

<u>Emergency core cooling system (ECCS)</u> - A backup system designed to supply cooling water to the reactor core in a loss-of-coolant accident.

<u>Emergency feedwater pumps</u> - Backup pumps intended to supply feedwater to the steam generators should the feedwater system fail to supply water. Also called auxiliary feedwater pumps.

<u>Feedwater pumps</u> - Two large pumps capable of supplying TMI-2's two steam generators with up to 15,500 gallons of water a minute.

GLOSSARY

Feedwater system - Water supply to the steam generators in a pressurized water reactor that is converted to steam to drive turbines; part of the secondary loop.

Fission - The splitting apart of a heavy atomic nucleus, into two or more parts when a neutron strikes the nucleus. The splitting releases a large amount of energy.

Fission products - Radioactive nuclei and elements formed by the fission of heavy elements.

Fuel damage - The failure of fuel rods and the release of the radioactive fission products trapped inside them. Fuel damage can occur without a melting of the reactor's uranium.

Fuel melt - The melting of some of the uranium oxide fuel inside a reactor.

Fuel rod - A tube containing fuel for a nuclear reactor.

Gamma rays - High-energy electromatic radiation; a form of ionizing radiation, of higher energy than X-rays, that penetrates very deep into body tissues.

General emergency - Declared by the utility when an incident at a nuclear power plant poses a potentially serious threat of radiation releases that could affect the general public.

General Public Utilities Corporation (GPU) - A utility holding company; parent corporation of the three companies that own TMI.

Genetic defects - Health defects inherited by a child from the mother and/or father.

Half-life - The time required for half of a given radioactive substance to decay.

Health physics - The practice of protecting humans and their environment from the possible hazards of radiation.

High pressure injection (HPI) - A pump system, capable of pumping up to about 1,000 gallons a minute into the reactor coolant system; part of the emergency core cooling system.

Iodine-131 - A radioactive form of iodine, with a half-life of 8.1 days, that can be absorbed by the human thyroid if inhaled or ingested and cause non-cancerous or cancerous growths.

Ionizing radiation - Radiation capable of displacing electrons from atoms; the process produces electrically charged atoms or ions. Forms include gamma rays, X-rays, and beta particles.

Isolation - Condition intended to contain radioactive materials released in a nuclear accident inside the containment building.

GLOSSARY

Krypton-85 - A radioactive noble gas, with a half-life of 10.7 years, that is not absorbed by body tissues and is soon eliminated by the body if inhaled or ingested.

Let-down system - A means of removing water from the reactor coolant system.

Loss-of-coolant accident (LOCA) - An accident involving a broken pipe, stuck-open valve, or other leak in the reactor coolant system that results in a loss of the water cooling the reactor core.

Make-up system - A means of adding water to the reactor coolant system during normal operation.

Make-up tank - A storage tank in the auxiliary building which provides water for the make-up pumps.

Meltdown - The melting of fuel in a nuclear reactor after the loss of coolant water. If a significant portion of the fuel should melt, the molten fuel could melt through the reactor vessel and release large quantities of radioactive materials into the containment building.

Metropolitan Edison Company (Met Ed) - Operator and part owner of the Three Mile Island nuclear power plant.

Millirem - 1 one-thousandth of a rem; see rem.

Natural cooling - The circulation of water without pumping by heating water in the core and cooling it in the steam generator.

Neutron - An uncharged particle found in the nucleus of every atom heavier than ordinary hydrogen; neutrons sustain the fission chain reaction in nuclear reactors.

Noble gases - Inert gases that do not react chemically and are not absorbed by body tissues, although they may enter the blood if inhaled into the lungs. These gases include helium, neon, krypton, xenon, and radon.

Nuclear Regulatory Commission (NRC) - U.S. agency responsible for the licensing and regulation of commercial, test, and research nuclear reactors.

Nucleus - The central core of an atom.

Pennsylvania Emergency Management Agency (PEMA) - Agency responsible for the state's response to natural and human-made disasters.

Person-rems - See collective dose.

"Poisons" - Materials that strongly absorb neutrons; used to control or stop the fission reaction in a nuclear reactor.

GLOSSARY

Pilot-operated relief valve (PORV) - A valve on the TMI-2 pressurizer, designed to open when steam pressure reaches 2,255 pounds per square inch.

Potassium iodide - A chemical that readily enters the thyroid gland when ingested. If taken in a sufficient quantity prior to exposure to radioactive iodine, it can prevent the thyroid from absorbing any of the potentially harmful radioactive iodine-131.

Pressure vessel - See reactor vessel.

Pressurizer - A tank that maintains the proper reactor coolant pressure in a pressurized water reactor.

Pressurized water reactor - A nuclear reactor system in which reactor coolant water is kept under high pressure to keep it from boiling into steam.

Primary system - See reactor coolant system.

Radioactive decay - The spontaneous process by which an unstable radioactive nucleus releases energy or particles to become stable.

Radioactivity - The spontaneous decay of an unstable atom. During the decay process, ionizing radiation is usually given off.

Radiolysis - The breaking apart of a molecule by radiation, such as the splitting of water into hydrogen and oxygen.

Reactor (nuclear) - A device in which a fission chain reaction can be initiated, maintained, and controlled.

Reactor coolant pump - One of four large pumps used to circulate the water cooling the core of the TMI-2 reactor.

Reactor coolant system - Water that cools the reactor core and carries away heat. Also called the primary loop.

Reactor vessel - The steel tank containing the reactor core; also called the pressure vessel.

Rem - A standard unit of radiation dose. Frequently radiation dose is measured in millirems for low-level radiation; 1,000 millirems equal one rem.

Respirator - A breathing mask that filters the air to protect against the inhalation of radioactive materials.

Safety-related - The NRC employs several broad definitions for this concept. By one, safety-related items are "structures, systems and components that prevent or mitigate the consequences of postulated accidents that could cause undue risk to the health and safety of the public." However, the NRC has no specific list of safety-related items. The licensee designates what in its plant is considered

GLOSSARY

safety-related. If the NRC disagrees, the question is negotiated. Safety-related items receive closer quality control and assurance, maintenance, and NRC inspection.

<u>Saturation temperature</u> - The temperature at which water at a given pressure will boil; the saturation point of water at sea-level is 212° F.

<u>Scram</u> - The rapid shutdown of a nuclear reactor, by dropping control rods into the core to halt fission.

<u>Secondary system</u> - See <u>feedwater system</u>.

<u>Site emergency</u> - Declared by the utility when an incident at a nuclear power plant threatens the uncontrolled release of radioactivity into the immediate area of the plant.

<u>Solid system</u> - A condition in which the entire reactor coolant system, including the pressurizer, is filled with water.

<u>Steam generator</u> - A heat exchanger in which reactor coolant water flowing through tubes heats the feedwater to produce steam.

<u>Steam table</u> - A chart used to determine the temperature at which water will boil at a given pressure.

<u>Teratogenesis</u> - The process of the development of gross abnormalities in the developing unborn child; these abnormalities or birth defects are not inherited.

<u>Thermoluminescent dosimeter (TLD)</u> - A device to measure nuclear radiation.

<u>TMI</u> - Three Mile Island; site of two nuclear power reactors operated by Metropolitan Edison Company.

<u>Transient</u> - An abnormal condition or event in a nuclear power system.

<u>Trip</u> - A sudden shutdown of a piece of machinery.

<u>Turbine building</u> - A structure housing the steam turbine, generator, and much of the feedwater system.

<u>Uranium Oxide (UO_2)</u> - A chemical compound containing uranium and oxygen that is used as a fuel in nuclear reactors.

<u>Waste gas decay tank</u> - One of two auxiliary building tanks in which radioactive gases removed from the reactor coolant are stored.

<u>Xenon-133</u> - A radioactive noble gas, with a half-life of 5.3 days, that is not absorbed by body tissues and is soon eliminated by the body if inhaled or ingested.

<u>Zircaloy-4</u> - A zirconium alloy from which fuel rod cladding is made.

PHOTO CREDITS: Allied Pix Service, Inc., p. 136; Courtesy of U.S. Department of Energy/EG&G, p. 108; Maps, Inc., p. 6; Courtesy of Metropolitan Edison Company, pp. 82, 84, 85, 92, 97, 140; Bill Pierce/CONTACT-CAMP, cover; United Press International Photos, pp. 121, 139; Wide World Photos, pp. 80, 95, 98, 105, 110, 113, 117, 127, 128, 132.

SUPPLEMENTAL VIEWS

BY MEMBERS OF

PRESIDENT'S COMMISSION ON THE

ACCIDENT AT THREE MILE ISLAND

Supplemental View by Six Commissioners

        Bruce Babbitt

        Carolyn D. Lewis

        Paul A. Marks

        Harry C. McPherson, Jr.

        Russell W. Peterson

        Theodore B. Taylor

Supplemental View by Bruce Babbitt

Supplemental View by John G. Kemeny

Supplemental View by Russell W. Peterson

Supplemental View by Thomas H. Pigford

Supplemental View by Anne D. Trunk

SUPPLEMENTAL VIEW BY SIX COMMISSIONERS

The Commission has unanimously recommended that: "In order to provide an added contribution to safety, the NRC should be required to the maximum feasible extent to site new power plants in locations remote from concentrations of population. Siting determinations should be based on technical assessments of various classes of accidents which can take place, including those involving releases of low dosages of radiation."

The undersigned six Commissioners voted for and support the following recommendation: "No new limited work authorization permits or construction permits should be issued until such time as the NRC or its successor has adopted siting guidelines consistent with the above recommendation."

Bruce Babbitt
Carolyn D. Lewis
Paul A. Marks
Harry C. McPherson, Jr.
Russell W. Peterson
Theodore B. Taylor

October 22, 1979

SUPPLEMENTAL VIEW BY COMMISSIONER BABBITT

It is with some misgiving that I feel compelled to add separate views to the report, for I find it to be a strong and lucid piece of work in almost every respect. Yet there are two areas where I feel the Commission stopped short of providing meaningful recommendations.

The most serious unresolved issue, in my opinion, of the entire inquiry is: Who should be allowed to run nuclear power plants?

A careful review of the the Commission findings and conclusions, along with the technical and legal staff reports upon which these are based, readily demonstrate that the utility in charge at Three Mile Island was not qualified to do and was not doing an adequate job. The record includes a listing of failures and inadequacies from maintenance to management, from operator's training to a lack of nuclear expertise at higher management levels. Our own findings state that "Met Ed did not have sufficient knowledge, expertise and personnel to operate the plant or maintain it adequately," and that "as a result of these deficiencies the safe operation of the TMI-2 plant was impaired."

This is a far reaching indictment of the utility in charge, the entity given the responsibility for controlling 15 billion curies of radioactivity. By the nature of its charge, the Commission explored in depth the operation capability and performance of just one nuclear utility and found it seriously wanting. But there are many indications that Met Ed is not an aberration, and that there are other nuclear utilities that do not measure up to even minimal standards. Inevitably, this raises serious questions about who should be licensed and entrusted to run our nuclear power plants. In my view, nuclear power is far too complex and dangerous to be left to any utility that wants it -- which has been the case until now. Nor can we allow utilities to go through a learning process at the expense of the public.

As a Commission, we had a real problem coming to grips with this issue because of the time constraints on examining the characteristics of other utilities operating nuclear power plants. I can, therefore, understand the difficulties in formulating a specific recommendation at this time.

Yet I must believe that our findings do support more than what we have said here by way of recommendations. We cannot simply urge the utility, industry, and the Nuclear Regulatory Commission to pay more attention to safety and to establish higher standards.

While this Commission has clearly addressed the institutional shortcomings of the NRC in its recommendations, it has not addressed the institutional problems of the industry.

Met Ed's operating license stems from an unquestioned assumption by the NRC, until now, that any utility that wanted to produce nuclear power could do so -- a policy that no matter how small or unsophisticated the utility, it was eventually entitled to wrap its arms around a nuclear reactor. Nuclear technology continues to proliferate throughout the industry with some forty utilities now operating reactors and with many more waiting in the wings.

There is no question that the management quality of utilities varies much more -- from very good to very mediocre -- than other major industrial sectors such as large chemical companies or computer manufacturers. And because utilities are necessarily monopolistic in nature, normal laws of competition do not apply; badly managed utilities suffer financial problems but somehow survive.

It is now time to assess this situation and determine which companies are qualified to handle such a technology and which companies are not. It is remarkable that this issue has not been previously confronted, but it is again a product of the "accidents can't happen" syndrome. Discriminating the good from the mediocre, the nuclear goats from the nuclear sheep, however unpalatable to the industry, must be done. One well known nuclear expert, Dr. Alvin Weinberg, has argued persuasively that the generation of nuclear power should be completely separated from the distribution of electricity and entrusted to just a few sophisticated entities with both the resources and the organizational depth to provide safe nuclear energy as their only task.

I believe that this is one area where fewer entities with more depth and expertise might be justified for the sake of public health and safety. Precisely how to control this proliferation of nuclear power management should receive a lot more study and I strongly urge the appropriate over-sight committees to place this issue near the top of their agenda.

Second, the Commission with its limited time and resources did not pursue in detail the issue of whether facts, known by Met Ed on the first day of the accident, were not communicated to NRC and state officials.

It now appears there is evidence to indicate that Met Ed technicians understood, within a few hours of the accident, that the nuclear core had been uncovered and that this specific information was transmitted to supervisory personnel at the plant early Wednesday. There seems to be little question that the technicians who took the temperature readings that morning understood what they found. The real question is what happened to this information and whether it was transmitted to the appropriate management personnel. It certainly did not get transmitted to responsible public officials, including Lt. Governor Scranton during a meeting with Met Ed that afternoon.

This incident again demonstrates the total inadequacy of the utility's internal communication system and raises serious questions about crisis management. As a Governor, it seems to me beyond question that a responsible public official must have immediate access to all available information about the status of a nuclear accident.

There is no question that this information might have influenced state and federal concerns over the need for evacuation then and subsequently. Whether or not an evacuation should have been ordered on the basis of the evidence known at the time is not particularly relevant now but the fact of the matter is that key decision-makers - those responsible for the public health and safety of the citizens - did not have access to the information that was known to the utility.

This issue should be intensely scrutinized by other investigatory bodies continuing the inquiry into nuclear power and this accident.

There are still unresolved questions about what happened at Three Mile Island, the answers to these may well lead to other recommendations about the responsibilities of utilities operating nuclear reactors.

Bruce Babbitt

October 25, 1979

SUPPLEMENTAL VIEW BY COMMISSIONER KEMENY

The Commission considered three different possible recommendations for a temporary halt on construction permits. Eight different Commissioners voted for at least one of these proposals. Unfortunately, we could not agree on the appropriate criteria for such a halt. Our reasons for failure to reach agreement are complex and may be found by examining the transcripts of our meetings of October 16, 20 and 21.

The following proposed recommendation was discussed extensively by the Commission:

"No new construction permits should be issued until the reports and recommendations of this Commission, the NRC self-evaluation and the Congressional investigations are complete and until the President and Congress have had an adequate opportunity to consider such recommendations, including the recommendation to restructure the NRC."

I was one of six Commissioners who voted in favor of this recommendation; four voted against it and two abstained. I very much regret that this important recommendation failed to obtain the seven-vote majority necessary to adopt it.

I was also one of four Commissioners who voted for a stronger version of the above recommendation.

John G. Kemeny

October 25, 1979

SUPPLEMENTAL VIEW BY COMMISSIONER PETERSON

Although I believe that our report fulfills well the President's charge and believe that our recommendations, if they were carried out, would reduce the likelihood of accidents, I wish to comment on the work of the Commission in three areas:

I. The Commission failed to summon the 7 votes necessary to adopt the following two resolutions:

A. "No new construction permits should be issued until the reports of this Commission, the NRC self-evaluation, and the Congressional investigations are completed and until the President and Congress have had an adequate opportunity to consider such recommendations including restructuring the NRC."

Six of the ten Commissioners who voted supported this resolution.

B. "No new limited work authorization permits or construction permits should be issued by the present NRC or the restructured NRC that are inconsistent with the siting recommendations in 6 and 6a."

(This reference is to approved recommendations that call for requiring, to the maximum feasible extent, the siting of new power plants in locations remote from concentrations of population.)

Six of the nine Commissioners who voted supported this.

In view of the strong support in our Commission for these two measures, I recommend that the Congress and the President enact them.

A minority within the Commission strongly resisted recommendations that might delay further nuclear plant construction. Neither the Commission nor its staff was free from the mind-set that nuclear energy is adequately safe--the mind-set for which the Commission criticized the NRC and the nuclear industry.

II. The study was not subjected to the penetrating critique which could have been provided by one or more of the highly technically qualified critics of nuclear energy safety available in our country. I recommend that the President and the Congress involve such experts in the continuing appraisal of the safety of nuclear energy. This is especially important when considering the possible accident conditions which can lead to a major release of radioactive material from the plant.

III. The Commission ruled that an investigation of the disposal of the TMI-2 nuclear wastes lay outside its assignment. Yet, in my view, this constitutes, over the long run, the most hazardous aspect of the nuclear power industry. While the industry waits for the government to finish its decades-long effort to determine how to safely dispose of these long-lived wastes such as plutonium, cesium and strontium, each nuclear power plant continues to store its growing amount of spent fuel containing these wastes in a pool of water immediately adjacent to the containment building.

I recommend that a serious study be undertaken of how such storage may exacerbate the threat from accidents or sabotage and of whether or not such waste should be moved away from the power plants, especially when the plant is located in a heavily populated area.

Although there is no commercial plant today for reprocessing spent fuel and our government refuses to approve one, the accident at TMI-2 has in effect converted that plant to a reprocessing plant. A large-scale chemical processing plant is being built at TMI-2 for handling the huge quantities of highly radioactive waste that have escaped from the disintegrated fuel rods. The safe processing and disposal of these wastes merit prompt and close surveillance by some independent group.

As a final comment, I wish to emphasize my conviction, strongly reinforced by this investigation, that the complexity of a nuclear plant--coupled with the normal shortcomings of human beings so well illustrated in the TMI accident--will lead to a much more serious accident somewhere, sometime. The unprecedented worldwide fear and concern caused by the TMI-2 "near-miss" foretell the probable reaction to an accident where a major release of radioactivity occurs over a wide area. It appears essential to provide humanity with alternate choices of energy supply. Accordingly, I recommend the development by our federal government, before we become more fully committed to the vulnerable nuclear energy path, of a strategy which does not require nuclear fission energy.

Russell W. Peterson

October 25, 1979

SUPPLEMENTAL VIEW BY COMMISSIONER PIGFORD

I generally concur with the conclusions and recommendations of the Presidential Commission on the Accident at Three Mile Island. However, some of the principal results of this investigation need clarification and discussion. Among these are some that warrant immediate, but necessarily limited, comment.

1. ### The Performance of Equipment and Engineering Systems

The Commission has properly recognized that, with the very heavy emphasis upon equipment to attain reactor safety, there has been too little emphasis upon the adequacy of people to help achieve that safety. The lack of such people emphasis has been properly stressed in this report. However, that stress has now obscured the very important fact that, in spite of the very crucial errors of operators and supervisors at TMI-2, the safety equipment did indeed function. In spite of the open PORV, leaks in the vent gas system, and other equipment failures, the overall system of equipment was sufficiently good that, without the human errors, the accident at TMI-2 would have been only a minor accident.

The reactor containment and its auxiliary equipment did indeed function to protect the public. Except for the small fraction that escaped to the environment, the radioactivity was contained. The off-site radiation doses were small. We have found that the actual release of radioactivity to the atmosphere will have a negligible effect on the physical health of individuals. Equipment failures were not the proximate cause of the TMI-2 accident. The accident was, in fact, a demonstration that the equipment *is* effective.

Although there has been considerable speculation about how near TMI-2 came to a worse accident, our staff analyses show that even if all of the reactor fuel cladding had been oxidized to form hydrogen, or even if appreciable fuel melting or even a meltdown had occurred, the containment would still have survived and protected the public. The accident demonstrated that the "defense-in-depth" approach towards nuclear reactor safety has indeed yielded significant results.

The emphasis in this report upon equipment vs. people obscures the fact that the equipment itself is only one product of the defense-in-depth or multiple-barrier design approach, which also encompasses the analysis of how equipment components must perform and how systems of equipment must operate. The accident demonstrated that this system of equipment performed better than expected. Earlier assumptions and studies by AEC/NRC (TID-14844 and WASH-1400) have suggested far greater core damage and greater releases of radioactivity from the fuel and into the containment under such degraded cooling conditions.

The accident has also demonstrated many areas wherein equipment modifications can result in further improvements in safety of existing and future reactors in this country.

These are important positive results from our investigation.

2. The People-Related Problem

The nature of the people-related problems needs clarification. One such problem--and a most serious one--was the errors made by operators and operator-supervisors, whose training was insufficient in scope and understanding. Another was the failure of many individuals to respond adequately to the earlier experience from other reactors and to other advance information that might have alerted the operators and avoided the accident.

Another problem was the errors made by some NRC officials, who misinterpreted the release of radioactivity on March 30 and recommended evacuation, and who erroneously concluded on March 31 that the hydrogen bubble might explode. The public trauma from these mistakes resulted in severe but short lived mental stress, which was evidently the only serious health effect of the accident.

Having identified the particular people-problems involved, many of the necessary direct remedies are apparent. There seems to be some unwillingness to recognize that many of these remedies are already being implemented. The NRC and the nuclear industry have taken and are taking steps on a broad basis to analyze and rectify these problems, as evidenced by the post-TMI NRC bulletins and by the establishment of the utilities' Institute for Nuclear Power Operations (INPO) and the reinsurance program. After experiencing the shock and comprehending the cost of this accident, the nuclear industry has clearly set into motion programs to institute many of the remedies that this Commission seeks. The problem with "attitudes" emphasized in the Commission's report must refer largely to pre-TMI attitudes.

It is reasonable to expect that other such human-related problems, not uncovered by this investigation, may exist. That, and the need to instill and continue a strong emphasis upon reactor safety, suggest some of the broader institutional changes recommended in this study.

3. Scope and Limits of the Investigation

The limits of this investigation and the effect thereof upon the Commission conclusions and recommendations need clarification.

This investigation was limited to the accident at TMI-2, and possible variations thereto, and, to a limited extent, similar transients at other places. The many other aspects of reactor safety were not investigated, although we do recommend that these be more systematically studied. The facts of the present investigation provide no basis for concluding that reactors are unsafe. They also show that, although more emphasis is needed on the analysis and planning for small-break accidents, the possibility of an accident of this type was known and had been analyzed and predicted prior to the TMI-2 accident. Therefore, any conclusions as to new fears of reactor safety do not arise from, and imply large extrapolations from, the facts of this investigation.

This investigation has not included a study of reactor siting. Consideration of the calculated "low population zone" occurred only in our consideration of its implication on the specification of radiation doses for evacuation decisions. Therefore, proposals made by some Commissioners to reverse existing site approvals in favor of more remote sites have no justification with the facts of this study.

We have recognized in this investigation that decisions as to whether or not safety improvements are to be implemented must be based in part upon a weighing of the costs against the benefits. However, we did not evaluate the costs of possible safety modifications, nor did we evaluate the probabilities of some of the large hypothetical releases that have been postulated by some Commissioners. Such proposals, and claims as to risks therefrom, have no basis within the facts of this investigation.

We have not investigated the availability, cost, overall safety, and environmental effects of nuclear energy and of other energy alternatives. Nor have we investigated the effect of various energy alternatives upon the nation's economy and security. We have not examined the effect of a speed-up or delay of nuclear power upon the many energy problems which affect the nation. Therefore, proposals by some Commissioners to impose sanctions which affect the availability of nuclear energy as an option are based upon their own personal extrapolations, which leap far beyond the facts of this investigation. The Commission, in its final consideration of the moratorium proposals, repudiated the issue by a vote of 8 to 4.

4. Lack of Input from Those Parts of the Nuclear Industry Not Involved Directly in TMI-2

Through its investigation of the Nuclear Regulatory Commission, the Commission staff has uncovered problems and practices which have suggested extrapolations to those many parts of the nuclear industry not involved directly with the TMI-2 accident. However, little proof of the validity of these extrapolations has been established. Moreover, to my knowledge, no representatives of those other parts of the nuclear industry were interrogated or asked to present evidence on any of the relevant issues, except for one company interrogated within the narrow issue of the Beznau incident. This further limits the validity of the industry-wide extrapolations that are implied in many places in the report and that are implied in some of the moratorium recommendations still endorsed by some of the Commissioners.

5. Attitudes

The framing of the Commission's overall conclusion around the question of

"attitudes of the Nuclear Regulatory Commission, and to the extent that the institutions that we investigated are typical, of the nuclear industry"

requires comment and interpretation. "Attitudes", especially prior to TMI-2, were not directly examined, nor could they be. Valid conclusions can only be drawn on actions taken, i.e., problems addressed and not addressed, regulations issued and complied with, and the occurrence of events that reflect upon the adequacy of those processes. Even if "attitudes" could be assessed, it is not clear how they could be changed by any recommended rule, reorganization, or other mandated influence. It is more constructive to assume that attitudes are symptomatic of the forces at work in the systems, and it is those forces which must be addressed.

The actions already taken by the industry in setting up INPO, the Nuclear Safety Analysis Center, and the program of self insurance against the cost of replacement power, with the self-policing actions thereby implied, signal a genuine, if somewhat belated, recognition of the need for greater effort to prevent nuclear accidents and cope with their consequences. These actions show a significant change in industry attitude which can only be beneficial.

It becomes clear, as the theme of "attitudes" is developed in the Commission report, that what is of concern is an apparent failure of the system to incorporate an effective mechanism to assimilate lessons from plant experience and to incorporate the appropriate up-to-date technology, particularly as it applies to control room design and to develop sufficiently trained and competent people to manage this technology. This is a more manageable and appropriate focus for the overall conclusion of this Commission.

I believe that such technology is being or will be used by the industry and that changes and improvements in design and operating procedure will be effected, not merely to satisfy critics nor to demonstrate attitudinal penitence, but on the basis of sound judgment resting on sound data.

6. Commission Judgments on Overall Safety

In its Overview the Commission acknowledges that it has not examined "how safe is safe enough or the broader question of nuclear vs. other forms of energy," recognizing the complexity of the issue and the limitations of staff. However, the Commission soon leaps this hurdle and speaks of the "risks that are inherently associated with nuclear power," and it holds that "equipment can and should be improved to add further safety." Even the conclusion that "accidents as serious as TMI should not be allowed to occur in the future" may imply that an assessment of risk and safety has been made. This conclusion is more understandable if interpreted in terms of what was really serious about this accident. The only serious health effect was the mental stress resulting from the confusion and public misunderstanding concerning the March 30 release and the March 31 hydrogen bubble. The financial loss to the utility and ultimately to the ratepayer is also serious.

Every technology imposes a finite degree of risk upon society, both in its routine operation and in the occurrence of accidents. Over a long enough time period, even low probability accidents may occur. The essential question is the trade-off between the risks and the benefits. The Commission neither received any evidence nor reached any conclusions that the risks of nuclear power outweigh its benefits.

7. *The NRC "Promotional Philosophy"*

The NRC's assignment is indeed difficult, but not because of dichotomy of safety, on the one hand, and the industry's convenience on the other. The problem is more complex. There is in each issue the element of how much cost, how many man-years of expert analysis, and how much delay is justifiable to achieve an increment of safety. Seldom are these issues black and white, since the designers and engineers must recognize that absolute absence of risk in any project is unattainable, and that social costs accrue to both inaction and overreaction. Efforts to balance costs and benefits should not be considered evidence per se of a promotional philosophy.

It should be expected that industry will logically resist unwarranted changes proposed in the name of safety.

8. *Hydrogen from Small-Break LOCAs*

Finding A-10 may be misinterpreted as suggesting that, because of the experience at TMI, the generation of large amounts of hydrogen gas is an inevitable consequence of small-break LOCAs. This misinterpretation leads to the erroneous conclusion that NRC over-emphasis on large-break LOCAs, at the expense of small breaks, is what left the TMI operators unprepared for the hydrogen produced during the accident, since significant amounts of hydrogen are not predicted in the typical analyses of large breaks. Such inference is without basis. Large-break analysis or any-break analysis will predict the generation of large amounts of hydrogen whenever the cooling water added to the reactor core from the emergency systems is reduced to the extent that was done at TMI-2.

9. *The Two-Step Licensing Process*

Finding G-6 implies that, in the two-step licensing process (Construction Permit and Operating License), safety may be compromised due to the large financial commitment prior to the operating license stage, with the implication that insufficient information is known at the construction permit stage for an in-depth safety review. A review of actual license applications will reveal that major safety features are sufficiently described at the construction permit stage. The issuance of an operating license several years later facilitates consideration of appropriate technological developments and feedback from operating plants which may be factored into the design toward the end of the construction period. Safety review in licensing is not a discrete two-step process. There is, and should be, continuing dialogue between the NRC staff and the applicant during this interim period.

10. *Single-Failure Criterion*

Finding G-8(a) that applicants "are not required to analyze what happens when two systems or components fail independently of each other" conveys some misunderstanding of the "single-failure" criterion. The requirement is that the applicant must show that applicable off-site radiation exposure limits will not be exceeded in the event of an accident initiated by:

(a) any credible component failure, and in which
(b) either all external or all internal power supply to the plant is lost, and

(c) there is, in addition, failure of that single active component whose failure would most worsen the results of the accident.

Although confusingly called a "single-failure" criterion, it is clear that this criterion requires the assumption of at least three failures.

It is further required that if failure of one component causes failure of other components, the entire series of failures must be regarded as one failure. The single-failure criterion is applied on a system-by-system basis, which implies single-failure tolerance in each of the systems.

11. Safety-Related

Finding G-8(b) concerning NRCs handling of "safety-related" items needs clarification in several respects. First, the well-established practice of the NRC is to require that any component, system, or feature needed for the prevention or mitigation of a serious accident must meet documented requirements of quality, redundancy, testability, environmental qualifications, etc., and must be categorized as "safety-related." Although other components, systems, or features are classed as "non-safety related" they must meet requirements appropriate to their operational function. NRC practice is to subject all "safety related" items to review. Additionally, "non-safety related" items are reviewed by NRC to reassess their possible reclassification.

Second, in analyzing postulated accidents, one is not permitted to assume that an active "non-safety related" item will be capable of performing its function. As a result, either an active item must meet "safety related" requirements of quality, etc., or no credit can be taken for its functioning in an accident.

In the TMI-2 accident it appears that the NRC's pre-occupation with the "safety-related" item list was not the fault, but rather the safety analyses did not take into account the actual lack of training, the inadequate operating procedures and practices, and their potential capability for producing an accident if the PORV stuck open.

Finally, the NRC is in some degree responsible for the level of safety consciousness in the industry. In this sense NRC's emphasis on "safety related" categories has probably been less influential than its reluctance to give credit for safety innovations and its requirement that the industry comply with many technically unreasonable rules. These practices encourage the industry merely to comply with NRC rules.

With regard to Finding G-8(c), it is not the reliance on "artificial categories of safety-related items" which has caused NRC to miss important safety problems. Rather, it was the failure to recognize that some items not part of the safety system may challenge that system at an undesirable frequency. Moreover, the capability of the operators to defeat the safety system was not given sufficient attention. These important issues are apart from safety-system classification and the single-failure criterion.

12. Plant Instrumentation

Finding G-8(f) does not provide a balanced account of all the considerations identified by AIF in its 1978 response to an NRC proposal to institute a new guide requiring a wider range of response for in-plant instrumentation, nor does it recognize the seeming lack of technical basis for the NRC request.

The relevance to the TMI-2 accident of the AIF response is not clear, since the range of the in-plant instrumentation at TMI-2 was adequate for diagnosis and plant control during the accident. Instead, the problem during the TMI-2 accident was that only part of the range of the in-plant instrumentation was displayed to the operators, and the manner of display was in some ways inadequate. Additionally, the operators misinterpreted some instrument readings. However, a greater range of instrument response might have aided the later assessment of the core damage that occurred.

13. Backfitting

Finding G-8(h), that there is no systematic backfitting review on a plant-by-plant basis of operating plants and plants under construction, appears to take too little account of the NRC's Systematic Evaluation Program (SEP), initiated more than three years ago. Under this program, operating plants have been categorized by NRC, issues have been identified by NRC, and information about older plants has been supplied to NRC by the utilities. In a number of cases, physical modifications of operating plants have been made in order to comply with updated NRC requirements. In some areas, such as that of the up-grading of emergency plans cited in the Commission's report, progress does appear to have been somewhat slow.

14. Independent Testing by I&E

In Finding G-9(a) and Recommendation 11(d) the recommended improvement of NRC's inspection and auditing of licensee compliance with regulations and the need for major and unannounced on-site inspections of particular power plants is logical. It calls for NRC to do more of what it already does and to do it better. In fact, NRC has, for over a year, stationed full-time inspectors at some operating nuclear power plants. At some plants, unannounced on-site inspections appear to be so frequent as to be commonplace.

The implication that NRC's I&E inspectors should do a substantial amount of independent testing of construction work and should place little reliance on work done by the utility is clearly impractical because of the enormous resources which would be required. Careful auditing of industry's testing is the only practicable and effective approach.

15. Emergency Procedures

In addition to the fact that some of the existing TMI-2 procedures were unworkable, as indicated in the Commission's report, the procedures did not provide a step-by-step pathway for identifying the problem implied by the information available in the control room. Given the philosophy that the operators had to adhere closely to written procedures, the unavailability of diagnostic procedures and training in their use was a significant factor among the causes of the TMI-2 accident.

16. The Major Problems with NRC's Approach to Reactor Safety

The Commission report has identified many mistakes by NRC personnel in their handling of the TMI-2 accident and deficiencies in NRC's regulatory practices. However, this criticism does not reach some essential elements of the problem. I believe that the following are some of the more important problems at NRC:

... Lack of quantified safety goals and objective. When a safety concern is postulated, there is no yardstick to judge the adequacy of mitigating measures.

... Inability to set priorities and to allocate resources in proportion to the estimated risk to the public. In my view, a disproportionate effort is being required for some issues which have only a marginal impact upon risk to the public.

... Lack of experienced staff. An undesirably large proportion of NRC staff and management have little or no practical experience in designing or operating the equipment which they regulate.

... Arbitrary requirements. Too many of the NRC requirements are mandated without valid technical back-up and value-impact analysis.

... A stifling adversary approach. The existing process inhibits the interchange of technical information between the NRC and industry. It discourages innovative engineering solutions.

... Ineffective evaluation of operations. NRC has no effective system for evaluating data from operating plants. Data should be analyzed systematically to identify trends and patterns.

... Lack of a comprehensive system approach to the whole plant. A large percentage of the NRC staff are specialists focusing upon narrow topics. There are relatively few systems engineers within NRC who can integrate individual safety features into an overall concept and who can place issues into perspective.

... An overwhelming emphasis on conservative models and assumptions. Realistic analyses are needed to identify the margins of safety and to aid competent decisions.

17. The Staff Report

The tight schedule and deadline for the Commissioners' report has allowed little opportunity for careful review of the Staff reports upon which our findings are to be based. Some Staff reports are not yet completed. There are several parts of some key Staff reports with which I cannot agree, particularly the staff report on the NRC.

18. The Staff Report on the Nuclear Regulatory Commission.

The Staff report on the Nuclear Regulatory Commission is a companion document published in Volume 2 of the Commission Report. Some deficiencies in this report are already reflected in earlier comments on Findings and Conclusions concerning the NRC. Having reviewed that report in search for understanding for many of the findings and conclusions adopted by this Commission, I noted several deficiencies, varying from technical error to unbalance in the investigation. Two examples are given below.

18.1 Performance Characteristics of Large Light-Water Reactors

The Staff report contains generalities by an NRC staff member, who seriously questioned the state of knowledge of the performance characteristics of the larger light-water reactors in this country, an opinion apparently also echoed by some other individuals within NRC. The cited statement was adopted by the authors of this Staff report. However, the Staff report reflects no attempt by the Staff to obtain evidence from the nuclear industry on this issue, even though the various companies in the nuclear industry are the parties impugned by the cited statements.

Statements were recently obtained from Saul Levine, Director of NRC's Office of Nuclear Regulatory Research, and from two different companies which design light-water reactors and which are not connected with the TMI-2 accident. It should not be construed from reference to "economy of scale" that the regulators were being asked to accept reduced safety margins. Rather, the growth was largely achieved by adding more fuel assemblies of the same or similar volumetric and linear power density, and by adding more heat transfer loops having the same mechanical and hydraulic characteristics as in the plants previously licensed. Saul Levine said, "as far as I know, there have been no size-dependent factors found in the operation of large reactors to affect the safety of the plants adversely." There appears no supportable suggestion that safety was compromised as a result of the extrapolation of technology.

The unqualified acceptance of the cited testimony in the Staff report is an indicator of insufficient balance in this part of the investigation.

18.2 Reliance on Books and Magazines

The Staff report relies to a considerable extent upon excerpts from a book authored by E. Rolph without establishing the author's qualifications. Ms. Rolph did not testify in this investigation. The undue reliance upon this secondary source, without first establishing a primary source for its support and without establishing its reliability, is a further example of insufficient balance in this part of the investigation.

In my view, the Rolph book does not express a comprehensive, accurate, and balanced knowledge of the NRC and of the nuclear industry.

19. Concluding Statement

The rather extensive criticism of NRC in the Commission report, and as implied in this supplementary statement, should not obscure the central issue that primary responsibility for nuclear safety lies with the utility, shared to a large extent with the equipment suppliers and the architect engineers. This also reflects my view of the responsibilities for the TMI-2 accident.

However, these criticisms of both industry and NRC should not obscure the fact that in 480 reactor years of commercial nuclear power operation in the United States there has still been no identifiable effect upon the physical health of the public, and that this record has been achieved by the industry and NRC, the parties that have been criticized and under the system that has been criticized.

It must be emphasized that nothing learned from this investigation suggests that the nuclear power option should be curtailed or abandoned as a result of the TMI-2 accident.

Thomas H. Pigford

October 25, 1979

SUPPLEMENTAL VIEW BY COMMISSIONER TRUNK

The following is a minority view on two issues raised in the report.

ITEM 1:

This item represents the feelings of the undersigned and a majority of her circle of citizens who lived through the TMI accident.

The report concluded that the errors and sensationalism reported by the news media merely reflected the confusion and ignorance of the facts by the official sources of information. It further concluded that the press did a creditable ("more reassuring than alarming") job of news coverage.

In fact, these conclusions are not generally supported by the staff reports. There were reliable news sources available. Too much emphasis was placed on the "what if" rather than the "what is." As a result, the public was pulled into a state of terror, of psychological stress. More so than any other normal source of news, the evening national news reports by the major networks proved to be the most depressing, the most terrifying. Confusion cannot explain away the mismanagement of a news event of this magnitude.

It is requested that the news media undertake a self evaluation on an individual basis and review their role in this accident which was not limited to equipment damage but also included psychological damage.

ITEM 2:

The undersigned could not support a motion for an undefined time frame moratorium on all new construction permits because it was not shown how this could result in a safer plant at TMI nor attain higher standards of safety and performance by the Industry.

A defined period (say two years) to act on this report's recommendations along with a separate probationary operating period (say five years) for the licensee at TMI could accomplish both the above objectives and is therefore recommended.

Anne D. Trunk

October 25, 1979